Hydrogen Infrastructure for Energy Applications

Hydrogen Infrastructure for Energy Applications

Production, Storage, Distribution and Safety

Hanane Dagdougui
Department of Mathematics and Industrial Engineering, Polytechnic School of Montreal, Canada

Roberto Sacile
Department of Informatics, Bioengineering, Robotics and Systems Engineering (DIBRIS), University of Genoa, Italy

Chiara Bersani
Department of Informatics, Bioengineering, Robotics and Systems Engineering (DIBRIS), University of Genoa, Italy

Ahmed Ouammi
Centre National pour la Recherche Scientifique et Technique (CNRST), Morocco

Academic Press is an imprint of Elsevier
125 London Wall, London EC2Y 5AS, United Kingdom
525 B Street, Suite 1800, San Diego, CA 92101-4495, United States
50 Hampshire Street, 5th Floor, Cambridge, MA 02139, United States
The Boulevard, Langford Lane, Kidlington, Oxford OX5 1GB, United Kingdom

© 2018 Elsevier Inc. All rights reserved.

No part of this publication may be reproduced or transmitted in any form or by any means, electronic or mechanical, including photocopying, recording, or any information storage and retrieval system, without permission in writing from the publisher. Details on how to seek permission, further information about the Publisher's permissions policies and our arrangements with organizations such as the Copyright Clearance Center and the Copyright Licensing Agency, can be found at our website: www.elsevier.com/permissions.

This book and the individual contributions contained in it are protected under copyright by the Publisher (other than as may be noted herein).

Notices

Knowledge and best practice in this field are constantly changing. As new research and experience broaden our understanding, changes in research methods, professional practices, or medical treatment may become necessary.

Practitioners and researchers must always rely on their own experience and knowledge in evaluating and using any information, methods, compounds, or experiments described herein. In using such information or methods they should be mindful of their own safety and the safety of others, including parties for whom they have a professional responsibility.

To the fullest extent of the law, neither the Publisher nor the authors, contributors, or editors, assume any liability for any injury and/or damage to persons or property as a matter of products liability, negligence or otherwise, or from any use or operation of any methods, products, instructions, or ideas contained in the material herein.

Library of Congress Cataloging-in-Publication Data

A catalog record for this book is available from the Library of Congress

British Library Cataloguing-in-Publication Data

A catalogue record for this book is available from the British Library

ISBN: 978-0-12-812036-1

For information on all Academic Press publications visit our website at https://www.elsevier.com/books-and-journals

Publisher: Joe Hayton
Acquisition Editor: Raquel Zanol
Editorial Project Manager: Mariana L. Kuhl
Production Project Manager: Vijayaraj Purushothaman
Cover Designer: Matthew Limbert

Typeset by SPi Global, India

Contents

Chapter 1

An Overview of Hydrogen Economy

1. OVERVIEW OF TODAY'S ENERGY SYSTEMS

The current economy reveals the continuing intensification in global energy consumption and carbon emissions. The gradual transition to alternative and renewable energy can be considered the most probable solution for the prevention of global warming intolerable and for the reduction of fossil fuels uses.

Fossil fuel-intensive development and urbanization have powered economic growth in many countries, but coal and oil combustion in power plants, industrial facilities, and vehicles represents the main cause of the outdoor pollution (IEA, 2016). Despite the increased attention given to renewable energy, the percentage of oil, coal, and natural gas in global energy consumption has been remarkably stable over the last 25 years: these fuels accounted for 81% of total energy use in 1989, exactly the same percentage as in 2014 (IEA World Energy Outlook, 2016). The main renewable energy source (RES) was hydropower and other renewable energy was only less than 2% of world primary energy consumption in 2012 (Hashimoto et al., 2016).

The widespread use of fossil fuels within the current energy infrastructure is considered as the major source of anthropogenic emissions of carbon dioxide, which is largely responsible for global warming and climate change (Balat, 2008). Currently, the level of CO_2 emissions per capita for developing nations is 20% of that for the major industrial nations. By 2030, CO_2 emissions from developing nations could account for more than half the world's CO_2 emissions. Another problem that is faced is the one related to the declining of crude oil supplied and political instability in the regions with large oil reserves. This dependency has developed the "energy security" concept and even deeper, the concept of "security of supply."

With regard to energy security, the EU has sought to address sovereignty, resilience, and robustness in the European Energy Security Strategy (EESS). The strategy defines energy security as a stable and abundant supply of energy, which in the long term should be provided in the context of a competitive, low-carbon economy that reduces the use of imported fossil fuels (European Commission, 2014).

Hydrogen Infrastructure for Energy Applications. https://doi.org/10.1016/B978-0-12-812036-1.00001-9
© 2018 Elsevier Inc. All rights reserved.

Worldwide energy consumption has been increasing rapidly and almost exponentially since the industrial revolution, and this increasing trend of energy consumption has been accelerated by the improvement of the quality of life that almost directly relates to the amount of energy consumption, by the industrialization of the developing nations, and by the increase of population in the world (Li, 2005). As a matter of fact, strict emission regulations are the subject of the worldwide discussion on the sustainability of the present fossil fuel-based energy systems. Furthermore, they are trigging the need to develop new energy systems driven by alternative fuels. There is an increasing need for new and greater sources of energy for future global energy and transportation applications. An alternative fuel must be environmentally acceptable, economically competitive, technically feasible, and readily available. The current trend of rising fossil fuel environmental-related problems and other energy security issues make the exploration for more sustainable ways to use energy more persuasive than ever. Sustainability is now a core concept in major policy-making processes, in the environmental aspects of population growth, and in the redevelopment of existing facilities and infrastructure. Sustainable development can be considered to be a high form of environmental policy. Due to the broad implications of resource management, sustainability is functionally dependent on energy use, on the ways energy is transformed, and the types of energy selected for a given task (Roosa, 2008).

The adoption of RES (renewable energy sources) and the rational use of energy, in general, are the fundamental inputs for any responsible energy policy. RES supply 14% of the total world energy demand (Demirbas, 2005).

According to (Observ'ER, 2013) the electricity production in the world from renewable sources including pumping stations reaches 4699.2 TWh in 2012, crossing the threshold of 20% of the world's electricity production. The renewable electricity production comes from six main sources: the first one is the hydroelectricity with a contribution in 2012 of 78%, followed by the wind energy that represents 11.4% of the total renewable production. While the biomass sector (solid, liquid, biogas, and waste), solar (PV, and thermal), and geothermal represent, respectively, 6.9%, 2.2%, and 1.5%.

RES can include biomass, hydropower, geothermal, solar, wind, and marine energies and are also often called green alternative sources of energy. The renewable are the primary, domestic, and clean or inexhaustible energy sources. Many authors from the academic and industrial worldwide communities agreed that RES are the pivot for sustainable development.

For instance, the European Union has officially recognized the need for promoting RES as a priority measure both for reduction of energetic dependence and for environmental protection (Franco and Salza, 2011). Achieving sustainable development on global scale will require the judicious use of resources, technology, appropriate economic incentives, and strategic planning at the local and national levels (Streimikiene et al., 2007).

A transition to a renewable-based energy system is crucial. Obstacles with renewable electric energy conversion systems are often referred to the intermittency and stochastic behavior of the energy sources. Their variability poses problems for applications that require a continuous supply of energy. Thus the variability related to the inherent temporal mismatch between source availability (sun shining, wind blowing, etc.) and the load poses a serious technical issue for the deployment of renewable energy (Bergen, 2008).

To summarize, the main critical elements related to the intermittent RES can be categorized as follows (Franco and Salza, 2011):

- Variability and stochastic behavior of renewable energy system due to fluctuation of renewable sources through weather conditions;
- Direct applications of energy from renewable can satisfy the electric loads but cannot be considered as an alternative fuel for transportation purposes, which means that electricity from RES cannot be useful in mobile applications;
- Constraint related to the flow of electricity sent from renewable energy systems through power grid lines;
- Mismatch between the supply and demand side, which make high pressure on servicing the demand side;
- Excess of electricity production risk, caused both by high level of installed renewable power plants capacity and by the intrinsic poor flexibility of thermal power plants.

In order to overcome the earlier RES drawbacks, one of the alternatives is the adoption of RES for hydrogen production. In fact, hydrogen is an energy carrier, so it needs to be produced from other sources of energy. A renewable-based hydrogen economy is one of the possible implementations of such systems. Using RES as basis of hydrogen production could lead to many environmental advantages. In addition, in contrast to RES alone, RES coupled with hydrogen production plants can solve many problems and can even increase the penetration of RES.

1.1 Toward Transition to Hydrogen Economy

The transition toward to "hydrogen economy" can be considered already started. Even if our economy is mostly based on fossil fuel, the technologies to produce hydrogen from hydrocarbons and water are ready. However, the industry sector has to accelerate the development of hydrogen energy infrastructures to deliver hydrogen to consumers in a clean, affordable, safe, and convenient manner.

Even when hydrogen utilization devices are ready for market applications, the consumers do not have the convenience to access hydrogen instead of gasoline, electricity, or natural gas today.

The main problem in the hydrogen economy diffusion is the high investment required for adapting the current infrastructures to hydrogen applications in several industry sectors including production, delivery, storage, conversion, and end-use technologies (Moreno-Benito et al., 2017).

Recently, extensive literature dedicated more attention to the hydrogen supply chain (HSC) optimization and design (Woo et al., 2016; Almansoori and Betancourt-Torcat, 2016; Kell and Hoyos, 2016). Dagdougui (2012) categorized literature on HSC into two types: (1) those related to a component of the supply chain, such as production, storage, distribution, market analysis and (2) those related to a complete analysis of the HSC including simultaneously all parts of the chain.

Currently, there are several topics which can affect hydrogen economy expansion.

The issues, which concern the global climate changes and the need to reduce greenhouse gas emission, encourage and support international and national institutions to promote hydrogen development. On the other side, the lack of incentives in the national energy policies, the complexity and the cost to realize hydrogen infrastructures, the negative public perception on hydrogen safety issues inhibit the hydrogen economy transition.

1.2 Conclusion

Energy is one of the principal driving forces in the economic, social, and technologic development of a country. It is the fundamental input for the production of goods and services, for increasing the health of the population, through the provision of thermal comfort, light, among other benefits (Pereira et al., 2008). Providing affordable, reliable, environmentally sustainable energy to the world's population presents a major challenge for the first half of this century and beyond. The sustainable development of RES requires a supply of energy that is sustainably available at a reasonable cost, without causing any environmental damages. It is well recognized that energy sources such as fossil fuels are finite and lack sustainability, while other alternatives like hydrogen are sustainable over a relatively longer term (Omer, 2008).

To realize significant progresses toward the adoption of a new hydrogen economy, Europe needs to promote and sustain energy policy actions that assess the notion that a hydrogen economy or hydrogen as an alternative fuel is an efficient and cost-effective resource of reducing dependence on fossil fuels and to tackle climate change challenges.

Probably, a strong public–private partnership on hydrogen energy development could be the essential component on the current economy to achieve a hydrogen vision. The process for a hydrogen economy transition required to identify, in a shared vision among industry, government, nongovernmental organizations, and research institutes, the near-, mid-, and long-term actions.

REFERENCES

Almansoori, A., Betancourt-Torcat, A., 2016. Design of optimization model for a hydrogen supply chain under emission constraints: a case study of Germany. Energy 111, 414–429.

Balat, M., 2008. Potential importance of hydrogen as a future solution to environmental and transportation problems. Int. J. Hydrog. Energy 33 (15), 4013–4029.

Bergen, A.P., 2008. Integration and Dynamics of a Renewable Regenerative Hydrogen Fuel Cell System. Ph.D. thesis, University of Victoria.

Dagdougui, H., 2012. Models, methods and approaches for the planning and design of the future hydrogen supply chain. Int. J. Hydrog. Energy 37 (6), 5318–5327.

Demirbas, A., 2005. Potential applications of renewable energy sources, biomass combustion problems in boiler power systems and combustion related environmental issues. Prog. Energy Combust. Sci. 31, 171–192.

European Commission, 2014. European Energy Security Strategy. COM(2014) 330 Final. European Commission, Brussels.

Franco, A., Salza, P., 2011. Strategies for optimal penetration of intermittent renewables in complex energy systems based on techno-operational objectives. Renew. Energy 36, 743–753.

Hashimoto, K., Kumagai, N., Izumiya, k., Takano, H., Shinomiya, H., Sasaki, y., Yoshida, T., Kato, Z., 2016. The use of renewable energy in the form of methane via electrolytic hydrogen generation using carbon dioxide as the feedstock. Appl. Surf. Sci. 388, 608–615.

IEA, 2016. World Energy Outlook 2016 Special Report Energy and Air Pollution—Executive Summary. http://www.iea.org.

IEA World Energy Outlook, 2016. Special Report. http://www.iea.org/publications/freepublications/publication/WorldEnergyOutlookSpecialReport2016EnergyandAirPollution.pdf. Accessed October 2016.

Kell, K.P., Hoyos, M., 2016. Modeling the Supply Chain Within Spain for Hydrogen Fuel Produced Using Sustainable Inputs. Dissertation.

Li, X., 2005. Diversification and localization of energy systems for sustainable development and energy security. Energy Policy 33, 2237–2243.

Moreno-Benito, M., Agnolucci, P., Papageorgiou, L.G., 2017. Towards a sustainable hydrogen economy: optimisation-based framework for hydrogen infrastructure development. Comput. Chem. Eng. 102, 110–127.

Observ'ER, 2013. http://www.energies-renouvelables.org/observ-er/html/inventaire/pdf/15e-inventaire-Chap01-Fr.pdf. Accessed October 2016.

Omer, A.M., 2008. Energy, environment and sustainable development. Renew. Sust. Energ. Rev. 12, 2265–2300.

Pereira, A.O., Soares, J.B., de Oliveira, R.G., de Queiroz, R.P., 2008. Energy in Brazil: toward sustainable development? Energ. Policy 36, 73–83.

Roosa, S.A., 2008. Sustainable Development Handbook. The Fairmont Press, Inc., Lilburn, GA.

Streimikiene, D., Ciegis, R., Grundey, D., 2007. Energy indicators for sustainable development in Baltic states. Renew. Sust. Energ. Rev. 11, 877–893.

Woo, Y.-b., Cho, S., Kim, J., Kim, B.S., 2016. Optimization-based approach for strategic design and operation of a biomass-to-hydrogen supply chain. Int. J. Hydrog. Energy 41 (12), 5405–5418.

Chapter 2

Hydrogen Production and Current Technologies

1. HYDROGEN AS AN ENERGY ALTERNATIVE

Global concern over environmental climate change linked to fossil fuel consumption has increased pressure to generate new clean energies. The exploitation of fossil fuels has created pollution on many local, regional, and global scales. In particular, the quality of air in big cities is in continuous deterioration due to gases emitted from transportation sector. Most of the scientific community, industrial, and government around the world are now convinced that the increase of carbon dioxide emissions and greenhouse gases in the atmosphere causes global warming with threatening consequences such as global climate change and sea level rise (Pearce, 2006; McCarthy et al., 2001). The transportation systems powered by hydrocarbons include increasing numbers of vehicles and are characterized by growing air pollution (Granovskii et al., 2006). In addition to the essential role of energy security that has motivated many countries worldwide to search for new sustainable energy alternatives, which are secure and environmental friendly. These concerns have arisen mainly in the world where energy demand is high and expected to grow rapidly over the coming century. Worldwide energy consumption has been increased rapidly (Hart et al., 1999), almost exponentially since the industrial revolution. This increasing trend has been also accelerated by the improvement of the quality of life that directly relates to the amount of energy consumption, by the industrialization of the developing nations. The shortage of energy will even put additional demands on global energy resources and expose an increasing number of countries of global economic activities to potential threats to supply security (Turton and Barreto, 2006). These facts persuade many countries to find alternative fuels that must be secure, environmentally acceptable, technically feasible, economically competitive, and readily available (Meher et al., 2006). The current fossil fuel systems must be changed gradually to clean and reliable energy systems, enabling to reach the sustainable vision of future energy systems. Among many alternatives, hydrogen can be considered as an attractive one to succeed the current carbon-based energy system (Ogden, 1999; Blanchette, 2008). Hydrogen is a secondary form of energy that has to be produced like electricity. It is an energy carrier. Its benefits are substantial

Hydrogen Infrastructure for Energy Applications. https://doi.org/10.1016/B978-0-12-812036-1.00002-0
© 2018 Elsevier Inc. All rights reserved.

considering the fact that hydrogen can be manufactured from a wide number of primary energy sources, such as natural gas, nuclear, coal, biomass, wind, and solar energies. In particular, it can be a key contribution to sustainable development, especially enhancing the penetration of renewable energy sources (RES) available around the world. The development of hydrogen infrastructures for producing and delivering hydrogen is primordial to reach the transition to a hydrogen economy. However, despite previous cited hydrogen advantages and benefits, the design of an upcoming hydrogen economy is a hard task. Hydrogen infrastructures with production facilities, distribution chains, and refilling stations are very expensive to construct. Real difficulties to hydrogen economy are the ones related to the need of significant investment costs with the no assurance of profitable demand, untitled as the "Chicken and Eggs" enigma (Waegel et al., 2006; Ogden et al., 2005).

In addition, from a demand viewpoint, the main obstacle in the adoption of hydrogen-fueled vehicles is the lack in delivery infrastructures. These problems motivate many scientists to study, understand, and analyze the hydrogen supply chain in advance, in order to detect the important factors that play major role in designing the optimal configuration. From the literature, studies related to hydrogen supply chain can be categorized into two types: (1) those related to a component of the supply chain, such as production, storage, distribution, market analysis and (2) those related to a complete analysis of the hydrogen supply chain including simultaneously all parts of the chain. The main objectives are (i) to review the studies conducted on the hydrogen supply chain; (ii) to classify many proposed approaches to find out the optimal configurations of the hydrogen supply chain, including the review of different available mathematical models developed to date to find the optimal configuration; and (iii) to present the future trends and challenges for the design of hydrogen supply chain.

2. HYDROGEN PRODUCTION METHODS

2.1 Hydrogen From Nuclear, Biomass, and Fossil Fuels

Nuclear power could produce hydrogen by either electrolysis of water or by direct thermal decomposition of water using heat from high temperature reactors (Moriarty and Honnery, 2007). Nuclear power plants produce heat that can be used directly or converted to electricity for the production of hydrogen. Hydrogen generation from water using nuclear energy has been examined in Japan. It is found that the high temperature gas cooled reactor (HTGR) has a possibility to generate hydrogen economically compared with other types of nuclear reactors. Four countries in the world are leaders in the production of hydrogen from nuclear: Japan, France, Korea, and United States. For instance, France is carrying out R&D program on massive hydrogen production with innovative high temperature processes. Collaboration in the literature appears also in the paper published by Yildiz and Kazimi (2006), who presented nuclear energy as major source for clean production of large amounts of hydrogen which will be essential for solving the problem of fast growing energy demand in all sectors in the world, including the transportation. Nevertheless, despite the R&D carried out in the world, the nuclear hydrogen production cannot be considered at any case as a sustainable route, especially after the explosion in the Fukushima Nuclear Power Plant triggered by the Tsunami in Japan 2011, thus nuclear faces many government and public acceptance.

Hydrogen produced from nonintermittent such as the biomass offers the possibility of renewable hydrogen. It enables a sustainable route for the production (Balat and Kırtay, 2010). The use of biomass instead of fossil fuels reduces the net amount of CO_2 released to the atmosphere. Biomass gasification can offer great potential through utilizing renewable feedstocks derived from agricultural waste, energy crops, and/or forestry residues (Guoxin and Hao, 2009). The gasification of biomass currently represents a global capacity of production of over 430 million Nm^3 of hydrogen per day (Shoko et al., 2006; Balat, 2008). Many countries around the world have allocated the research and development toward the hydrogen production from biomass. For instance, Austria has several demonstration plants and pilot projects for the gasification of biomass for hydrogen production. In the United States, a target objective is to lead to hydrogen from biomass competitive with gasoline by 2015. Despite the advantages of hydrogen from renewable biomass, biomass has several limitations; among them, the processes of hydrogen production from biomass are still in development and require a strong effort in terms of R&D and demonstration activities (Balat and Balat, 2009). In addition, another limitation illustrated in terms of the low content of hydrogen available in the biomass (6%–6.5%). Also, the characteristics of biomass are very important since they can vary greatly from location to location, seasonally, and yearly (Fowler et al., 2009). So that the hydrogen production via biomass route may not be competitive with the hydrogen production with fossil fuels. From other production points, the costs of hydrogen production from biomass become higher due to the additional costs of the logistics of the crops to the centralized production facilities (Landry, 2006).

Hydrogen can be manufactured from fossils fuels through a variety of technologies (coal, natural gas, etc.). The production technologies are steam methane reforming (SMR), partial oxidation, and gasification. Nowadays, most hydrogen produced worldwide (approximately 99%) which is about 700 billion Nm^3 per year is derived from fossil fuels—roughly half on natural gas and close to one-third on crude oil fractions in refineries. At the world scale, global hydrogen production today is enough to fuel approximately more than Nm^3 600 million fuel cell cars and is based almost exclusively on fossil fuels. Currently, the largest use of hydrogen is as a reactant in the chemical and petroleum industries: ammonia production has a share of around 50%, followed by crude oil processing with slightly less than 40%. In Europe, 80% of the total hydrogen was consumed by mainly two industrial sectors: the refinery (50%) and the ammonia industry (32%), which are both captive users. If one adds hydrogen consumption by methanol and metal industries, those four sectors cover 90% of the total. The hydrogen production from fossil fuels is a well-established technology in the world; for instance, in Europe, there are seven high capacity producers in Germany, United Kingdom, Netherlands, Spain, France, Belgium, and Italy.

2.2 Hydrogen From Renewable Energy Resources

Many authors agreed that neither fossil fuel, biomass, or nuclear cannot satisfy the existing electricity demands and cannot provide sufficient climate neutral energy to be probable routes for long-term hydrogen future production. Biomass, hydro, and geothermal even their feedstock can be estimated by accuracy, but they have limited potential and they are not always climate neutral (Moriarty and Honnery, 2007). The only remaining way to produce hydrogen is then the intermittent RES, especially solar and wind energies. In contrast to those production methods, renewable energies are a desired energy source for hydrogen production due to their diversity, region, abundance, and potential for sustainability. Renewable hydrogen is mainly an economic option in countries with a large renewable resource base and/or a lack of fossil resources, for remote and sparsely populated areas (such as islands) or for storing surplus electricity from intermittent renewable energies. Hydrogen production from renewable energy provides a great energy storage medium that can be exploited in the transportation sector, enabling then new interactions between the renewables and transportation sector. Renewable hydrogen production offers the potential for a distributed hydrogen supply network model, which would be based on on-site or off-site hydrogen production. The electricity generated from renewable sources can be turned into hydrogen using the electrolysis process. In fact, about 55 kWh of electricity are needed to liberate 1 kg of hydrogen from 9 kg of water by electrolysis (Bossel, 2006). Hydrogen can then be stored until it is needed as fuel for either transportation applications or local stationary power generation sources, or it can be transferred into electricity and fed into the electrical (Martin and Grasman, 2009). Electrolysis driven by renewable energy may be an option for a sustainable hydrogen production. In fact, the electricity generated by the renewable energy systems is transferred to the electrolyzers system for the production of hydrogen via electrolysis by passing electricity through two electrodes in water. One advantage of electrolysis of water is that nowadays it is compatible with a large variety of available renewable energy technologies, namely, solar, hydro, wind, wave, geothermal, etc. In addition, water electrolysis benefits of some additional advantages (Clarke et al., 2009), among them the use of different scales (on-site and off-site), its greater maturity, compactness and high current density, and small footprint. General interest in a wind–hydrogen system has increased partly because the price of wind power has become competitive with traditional power-generating

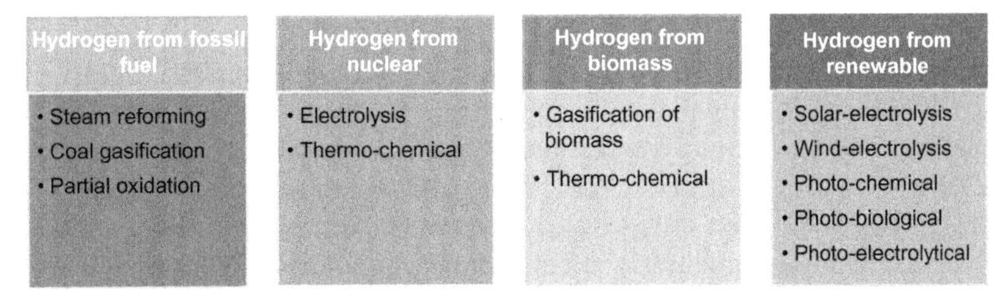

FIG. 2.1 Hydrogen production process.

sources in certain areas. Due to the characteristics of a wind-hydrogen system it has the potential to play a complementary role during the mass introduction of hydrogen. It seems likely that intermittent RES, chiefly wind and solar will have to supply most nonfossil energy in 2050 and beyond (Moriarty and Honnery, 2009). The technical potential for these sources is undoubtedly very large. In the long term, strong hydrogen markets and a growing hydrogen infrastructure will create opportunities for renewable hydrogen systems. To be successful, many challenges face the wind and solar energies, among them cost issues and the need for improvement in energy ratio (Moriarty and Honnery, 2007). Fig. 2.1 displays different technologies available for hydrogen production.

2.3 Comparisons of Hydrogen Production Routes

2.3.1 Scales of Production

The cost of hydrogen is highly influenced by the scale of the installation. In fact, in addition to the various production routes available for the generation of hydrogen, more pathways could be developed, namely, distributed and centralized production pathway. Centralized production is usually a large-scale production of hydrogen, where hydrogen is needed to be transported to the demand points. For instance, the hydrogen production cost from natural gas via steam reforming of methane varies from about 1.25 US$/kg for large systems to about 3.50 US$/kg for small systems with a natural gas price of 0.3 US$/kg. Distributed production is considered by many authors to be the most likely pathway during the market development of energy systems. In this case, hydrogen must be used close to the production point (Levin and Chahine, 2010). The distributed production infrastructure could consist of natural gas reformers or electrolyzers located at the point of use, for example, refueling station or stationary power generation. This pathway does not require substantial hydrogen delivery infrastructure. The cost of decentralized H_2 production may exceed US$6/kg today (hydrogen production and distribution). The centralized production benefits from large economies of scale, but to be commercially viable there is a need to develop distribution technologies.

2.3.2 Environmental Benefits and Challenges

The environmental benefits of the hydrogen production need to be well investigated in order to study the effects of different production means on the environment. The long-term hydrogen production must ensure an environmental sustainability so to be successful and to not lead to the same problems of today's conventional fossil fuels. Many papers have studied the environmental impacts of various hydrogen production routes, including those based on nonrenewable and renewable energy resources (Kothari et al., 2008; Afgan and Carvalho, 2004; Williams, 2004; Afgan and Carvalho, 2002). These impacts are always evaluated in terms of the carbon dioxide emitted for the production of 1 kg of hydrogen, since CO_2 is the most important greenhouse gas and is the largest emission from the systems (Kothari et al., 2004). In particular, steam reforming of methane (natural gas) requires only 4.5 kg of water for each kilogram of hydrogen, but 5.5 kg of CO_2 emerge from the process (Bossel, 2006). Fig. 2.2 displays the CO_2 emissions for different hydrogen production technologies:

SMR, steam methane reforming;
CG, coal gasification;
PV-EL, photovoltaic and electrolyzer;
H-EL, hydropower and electrolyzer;
POX, partial oxidation of hydrocarbons;
BG, biomass gasification;
W-EL, wind power and electrolyzer.

Hydrogen production from fossil fuels is a major source of CO_2 emissions. It appears that the main disadvantages of hydrogen production via SMR, coal gasification (CG), and partial oxidation of hydrocarbons (POH) are the emissions of the CO_2. Among H_2 production routes, coal gasification is the one that

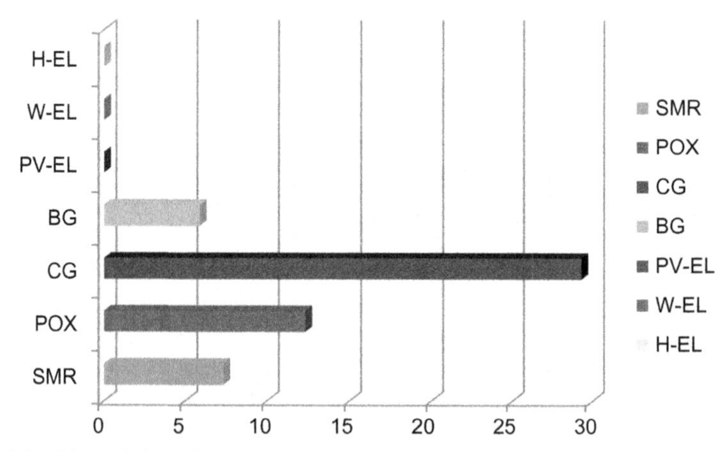

FIG. 2.2 CO_2 emissions of various hydrogen production technologies (Pilavachi et al., 2009).

lead to high emissions (equal to 29.33 kg CO_2/kg H_2). The electrolysis is the one considered as the only process that is not accompanied with the CO_2 emissions. It is considered so only when power plants use renewable energy resources to generate the needed electricity.

Some solutions are proposed to reduce the emissions of CO_2 such as the separation and the sequestration of carbon dioxide produced in the hydrogen production process capture. Nevertheless, the CO_2 sequestration is not yet technically and commercially proven, and the additional cost of the logistics of the CO_2 capture, storage, and transport could increase the total cost of hydrogen production from fossil fuel. As reported by Ball et al. (2007), the total hydrogen production costs increase by about 3%–5% in the case of natural gas reforming and 10%–15% in the case of coal gasification.

2.3.3 Costs and Economy of Hydrogen Production

To complete the comparison among the different processes, the cost analysis is important to determine whether a certain hydrogen production route can be used. The cost of producing hydrogen depends on the capital, operation, maintenance, and feedstock costs (Kothari et al., 2008). For instance, from feedstocks viewpoint, cost of hydrogen from fossil fuels is highly dependent on the price of natural gas and other conventional fuels, while the cost of hydrogen produced from renewable energy resources depends on the level of advancement of renewable energy technologies, and whether the system is connected or not to the electric grid. Figs. 2.3–2.5 show the previous costs for various production routes as analyzed by Pilavachi et al. (2009), while Table 2.1 presents the cost of hydrogen in \$/kg.

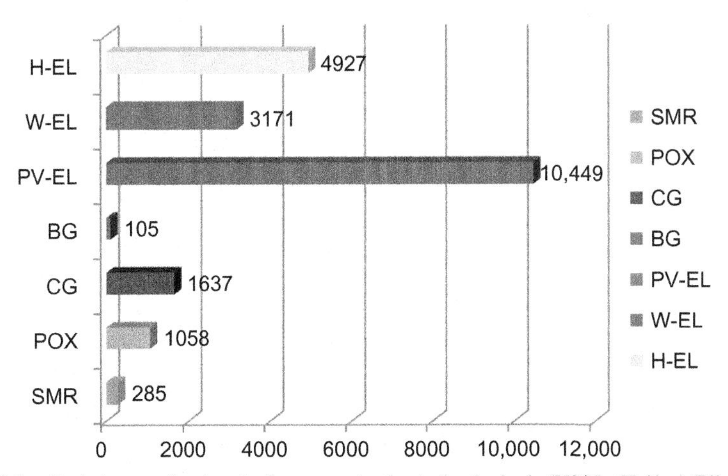

FIG. 2.3 Capital costs of various hydrogen production technologies in (US\$/kg H_2/day) (Pilavachi et al., 2009).

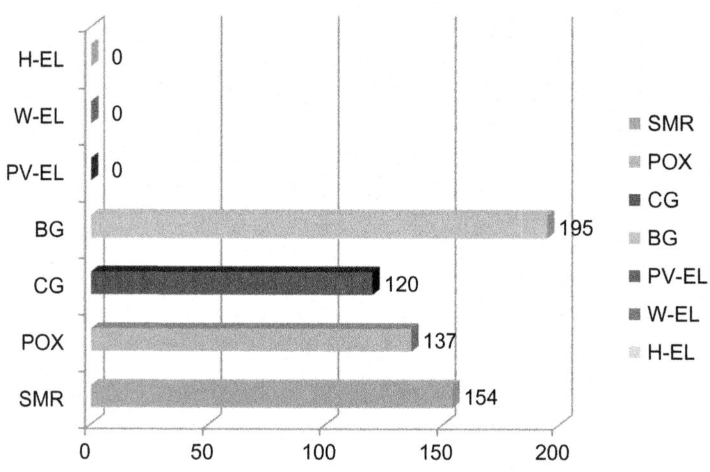

FIG. 2.4 Feedstock costs of various hydrogen production technologies in (US$/kg H_2/day) (Pilavachi et al., 2009).

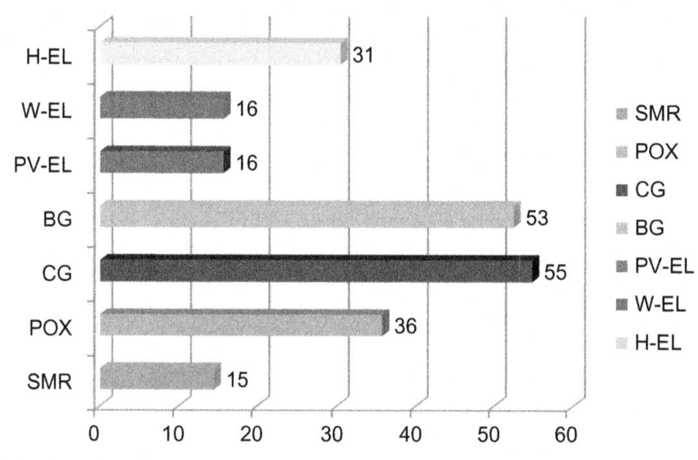

FIG. 2.5 Operating and maintenance costs of various hydrogen production technologies in (US$/kg H_2/day) (Pilavachi et al., 2009).

Fig. 2.4 displays that the biomass gasification technology is the one that have higher feedstock costs (followed by the SMR technology). No feedstocks cost is been added to the cost of hydrogen from renewable energy resources. In Table 2.1, the hydrogen production from renewable energy resources has higher capital cost, where a higher value (10,449 $/kg H_2/day) is observed for the electrolysis process with photovoltaic technology (PV-E).

From the last section, it appears that there are many trade-offs that exists among the production routes of hydrogen. From an environmental perspective,

TABLE 2.1 Cost of hydrogen from various technologies

Resources	Cost of hydrogen production ($/kg)
Methane steam reforming	0.828 (Mueller-Langer et al., 2007; Lagorse et al., 2008)
Nuclear	1.44–5.40 (FY 2008 Annual Progress Report, 2008; Elder and Allen, 2009)
Biomass	5.28–9.84 (Lemus and Martínez Duart, 2010) (biomass California)
Hydropower	5.4–7.92 (Mueller-Langer et al., 2007)
Geothermal	9 (Ewan and Allen, 2008)
Wind electrolysis	4–9 (Greiner et al., 2008)
PV electrolysis	5–20 (Lemus and Martínez Duart, 2010)

it is evident that hydrogen production from renewable energy resources, in particular, wind, hydro, and PV will have significant effect on the reduction of the carbon dioxide emissions. From an economic perspective, by comparing the different costs of hydrogen production methods, it appears that the same previous resources have no costs related to feedstocks since these resources are free and available. The operating and maintenance costs are still low by using the wind for hydrogen production, but once it comes to the capital costs, photovoltaic has higher capital costs, followed by wind turbines, and hydro. In our thesis, we have based our study using the RES giving special attention to wind and solar. These production sources will be implemented in the thesis as main feedstocks for the hydrogen supply chain. We believe that these resources could lead to high environmental benefits, as regarding the costs, they may go down as more technologies will be developed.

3. RENEWABLE HYDROGEN ENERGY

In order to overcome the earlier RES drawbacks, one of the alternatives is the adoption of RES-hydrogen systems. In fact, hydrogen is an energy carrier, so it needs to be produced from other sources of energy. A renewable-based hydrogen economy is one of the possible implementations of such systems. Using RES as basis of hydrogen production could lead to many environmental advantages. In addition, in contrast to RES alone, RES coupled with hydrogen production plants can solve many problems and can even increase the penetration of RES. This penetration could be reached through creating a bridge to the use of RES in the transportation sector and through the use of hydrogen as a storage medium for electricity generation. Hence, hydrogen can play double

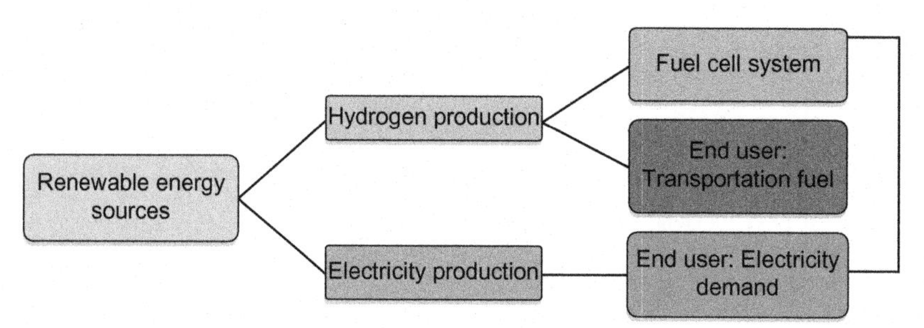

FIG. 2.6 Renewable energy possible uses with electricity and hydrogen.

role: as a fuel for the transportation sector and as a storage medium for the intermittent RES. The main question now is why hydrogen and not another material. In fact, hydrogen is the most abundant element in the world, it is not toxic and its combustion does not create any pollution or greenhouse gases. It has the highest specific energy content of all conventional fuels. Hydrogen may be used as fuel in almost any application where fossil fuels are used today, in particular in transportation sector, where it would offer immediate benefits in terms of the pollution reduction and cleaner environment (Barbir, 2009). Hydrogen, in contrast to RES, can be utilized in different applications and in all parts of the economy (e.g., generate electricity and as an automobile fuel). Recent studies suggest that renewable hydrogen will have a small price premium compared to hydrogen from traditional fossil sources (such as steam reforming of natural gas) if hydrogen is to become a competitive, environmentally friendly alternative to gasoline (NAS, 2004). In long term, hydrogen from renewable energy can contribute significantly as a bridge to the implementation of RES in transportation sector. In addition, hydrogen could be more used as a storage medium for electricity from intermittent renewable energies such as wind power and photovoltaic systems, contributing then significantly to the widespread use of renewable energy technologies. Fig. 2.6 displays the possible uses of RES. It appears clearly how the perspective of further increase of RES can lead to satisfy transportation and electricity demand.

3.1 Renewable Hydrogen: As a Future Fuel

Hydrogen is currently gaining much attention as a possible future substitute for fossil fuel in the transport sector (Lee et al., 2010). As transport fuel, hydrogen can both be combusted like kerosene and gasoline, or used in fuel cells. Fuel cells convert hydrogen into electricity, which is used to power the electric motors that are used to make the wheels turn and propel the vehicles. The use of hydrogen in fuel cell vehicles offers a number of advantages over existing fuels (Hugo et al., 2005). The fuel cell vehicles (FCV) can be a long-term solution to the persistent environmental problems associated with transportation.

FCV are less complex, have better fuel economy, lower GHG emissions, greater oil import reductions, and would lead to a sustainable transportation system once renewable energy was used to produce hydrogen (Thomas et al., 1998). The internal combustion engine can operate in a bivalent mode with gasoline and hydrogen from RES, which might be of significant importance for the transition period to a hydrogen economy in the early phase of market introduction (Ajanovic, 2008). According to Wang and Lin (2009)), six issues must be addressed in order to implement the hydrogen as an alternative fuel, namely, limited number of refueling station, high cost of installation of refueling stations, limited on-board fuel storage, safety and liability concerns, improvements in the competition, and high initial cost for consumers. There are a few ways to supply renewable energy to drive vehicles. One is to use electricity from a grid, which will be supplied by wind power or PV (photovoltaics). In this case electricity will be used by plug-in electricity or by hydrogen production through water electrolysis (Tsuchiya, 2008).

The option of producing hydrogen from RES as a transportation fuel is also receiving increased attention. Many studies have been published which investigate the technical and economic issues of hydrogen as a future fuel. Linnemann and Steinberger-Wilckens (2007) have presented a model that calculates realistic costs of wind–hydrogen vehicle fuel production. Authors found that if hydrogen is to represent a practical fuel alternative, it has to compete with conventional energy carriers. Reference Joffe et al. (2004) presented a technical modeling of hydrogen infrastructure technologies and how they could be deployed to provide an initial facility for the refueling of hydrogen fuel cell buses in London. The results suggest that the choice of H_2 production technology can have significant effects on when the infrastructure would be installed, and the timing of hydrogen production, and bus refueling. In reference Vidueira and Contreras (2003), a feasibility study of an autonomous solar-H_2 system connected to a filling station for fuel cell buses is presented. Greiner et al. (2008) gave a method for assessment of wind-hydrogen energy systems. Authors evaluate two possibilities with connection and no connection with the grid. The produced hydrogen will be supplied to a filling station where local vehicles can fill at demand.

3.2 Renewable Hydrogen: As an Electric Bridge

Hydrogen is proposed as a convenient energy carrier due to its versatility in use and as an energy storage medium. Hydrogen produced from RES offers the promise of a clean, sustainable energy carrier that can be produced from domestic energy resources around the globe. Energy storage is a potential solution to the integration issues that are described previously. Appropriate operation of energy storage could increase the value of wind and solar power in the power system through ensuring the matching between renewable power generation and electric demand. Hydrogen is now in the same position electricity was a little over century ago, when it replaced the direct use of the power from a steam engine (Scott, 2004). However, hydrogen does not replace electricity in the future, but they will work together in some kind of synergy. Electricity will be converted to hydrogen when energy storage is needed and hydrogen will be converted back to electricity when, e.g., a fuel cell vehicle needs power to its traction motor (Sperling and Cannon, 2004). The implementation of hydrogen will help to overcome the storage difficulty of renewable energy. A renewable hydrogen system with electrolyzer, storage, and fuel cell can be used to provide households with a reliable power supply. In fact, hydrogen has been regarded by many as a popular carrier of renewable energy in remote locations (Barbir, 2009; Wietschel and Seydel, 2007; Salgi et al., 2008). Hydrogen can be expected to allow the integration of some RES, of an intermittent character, in the current energy system. Several authors have proposed the use of hydrogen for the management of the energy generated. In fact, the electricity generated from renewable sources such as wind, solar, and hydro can be turned into hydrogen using the electrolysis process, hydrogen can then be stored until it can be transferred into electricity and fed into the electrical grid. Hydrogen is the suitable energy carrier to store solar and wind energy and transforms them to most convenient energy form of electricity. Electricity made from RES is an inexhaustible, environmentally friendly energy carrier. International interest in hydrogen as an energy carrier is high. Several projects have considered the feasibility of a renewable hydrogen economy, and others have begun to plan renewable hydrogen systems.

4. CONCLUSION

The increasingly environmental impacts of global warming have made a worldwide priority to phase out the use of fossil fuels as transportation and energy fuels in favor of hydrogen. The alternative fuels and energy carriers that are produced from the RES are challenging for the sustainable development of renewable energy. These systems need a lot of investigations to first manage the flux of renewable energy and then to produce the alternative fuel and energy. In addition, one important aspect that worth to be studied is the feasibility of such systems, and what are the limits of considering renewable energy as a source of fuel and electricity production. During the further development of the thesis, these aspects will be deeply studied, analyzing in the same time the aspects related to the feasibility of the green production routes and developing a supply chain that operates based on such clean renewable energy resources.

REFERENCES

Afgan, N.H., Carvalho, M.G., 2002. Multi-criteria assessment of new and renewable energy power plants. Energy 27, 739–755.

Afgan, N.H., Carvalho, M.G., 2004. Sustainability assessment of hydrogen energy systems. Int. J. Hydrog. Energy 29, 1327–1342.

Ajanovic, A., 2008. On the economics of hydrogen from renewable energy sources as an alternative fuel in transport sector in Austria. Int. J. Hydrog. Energy 33, 4223–4234.

Balat, B., 2008. Potential importance of hydrogen as a future solution to environmental and transportation problems. Int. J. Hydrog. Energy 33, 4013–4029.

Balat, H., Kırtay, E., 2010. Hydrogen from biomass—present scenario and future prospects. Int. J. Hydrog. Energy 35, 7416–7426.

Balat, M., Balat, M., 2009. Political, economic and environmental impacts of biomass-based hydrogen. Int. J. Hydrog. Energy 34, 3589–3603.

Ball, M., Wietschel, M., Rentz, O., 2007. Integration of a hydrogen economy into the German energy system: an optimising modelling approach. Int. J. Hydrog. Energy 32, 1355–1368.

Barbir, F., 2009. Transition to renewable energy systems with hydrogen as an energy carrier. Energy 34, 308–312.

Blanchette, S., 2008. A hydrogen economy and its impact on the world as we know it. Energ Policy 36, 522–530.

Bossel, U., 2006. Does a hydrogen economy make sense? Proc. IEEE 94 (10), 1826–1837.

Clarke, R.E., Giddey, S., Ciacchi, F.T., Badwal, S.P.S., Paul, B., Andrews, J., 2009. Direct coupling of an electrolyser to a solar PV system for generating hydrogen. Int. J. Hydrog. Energy 34, 2531–2542.

Elder, R., Allen, R., 2009. Nuclear heat for hydrogen production: coupling a very high/high temperature reactor to a hydrogen production plant. Prog. Nucl. Energy 51, 500–525.

Ewan, B.C.R., Allen, R.W.K., 2008. A figure of merit assessment of the routes to hydrogen. Int. J. Hydrog. Energy 30, 809–819.

Fowler, P., Krajacic, G., Loncar, D., Duic, N., 2009. Modelling the energy potential of biomass—H2RES. Int. J. Hydrog. Energy 34, 7027–7040.

FY 2008 Annual Progress Report, 2008. DOE Hydrogen Program. Available from: www.hydrogen.energy.gov/pdfs/progress08/ii_0_hydrogen_production_overview.pdf.

Granovskii, M., Dincer, I., Rosen, M.A., 2006. Environmental and economic aspects of hydrogen production and utilization in fuel cell vehicles. J. Power Sources 157, 411–421.

Greiner, C., Korpås, M., Holen, A., 2008. A Norwegian case study on the production of hydrogen from wind power. Int. J. Hydrog. Energy 32, 1500–1507.

Guoxin, H., Hao, H., 2009. Hydrogen rich fuel gas production by gasification of wet biomass using a CO_2 sorbent. Biomass Bioenergy 33, 899–906.

Hart, D., Freund, P., Smith, A., 1999. Hydrogen-Today and Tomorrow. IEA Greenhouse gas R&D Programme. ISBN: 1-898373-24-8. Available from: http://www.ieagreen.org.uk/h2rep.htm.

Hugo, A., Rutter, P., Pistikopoulos, S., Amorelli, A., Zoiac, G., 2005. Hydrogen infrastructure strategic planning using multi-objective optimization. Int. J. Hydrog. Energy 30, 1523–1534.

Joffe, D., Hart, D., Bauen, A., 2004. Modelling of hydrogen infrastructure for vehicle refueling in London. J. Power Sources 13, 13–22.

Kothari, R., Buddhi, D., Sawhney, R.L., 2004. Sources and technology for hydrogen production: a review. Int. J. Global Energy Issues 21, 154–178.

Kothari, R., Buddhi, D., Sawhney, R.L., 2008. Comparison of environmental and economic aspects of various hydrogen production methods. Renew. Sust. Energ. Rev. 12, 553–563.

Lagorse, J., Simões, M.G., Miraoui, A., Costerg, P., 2008. Energy cost analysis of a solar-hydrogen hybrid energy system for standalone applications. Int. J. Hydrog. Energy 33, 2871–2879.

Landry, B.M., 2006. Green Hydrogen: Site Selection Analysis for Potential Biomass Hydrogen Production Facility in the Texas-Louisiana Coastal Region. Master Thesis, Faculty of the Louisiana State University and Agricultural and Mechanical College.

Lee, Y., Kim, J., Kim, J., Kim, E.J., Kim, Y.G., Moon, I., 2010. Development of a web-based 3D virtual reality program for hydrogen station. Int. J. Hydrog. Energy, 1–7.

Lemus, R.G., Martínez Duart, J.M., 2010. Updated hydrogen production costs and parities for conventional and renewable technologies. Int. J. Hydrog. Energy 35, 3929–3936.

Levin, D.B., Chahine, R., 2010. Challenges for renewable hydrogen production from biomass. Int. J. Hydrog. Energy 35, 4962–4969.

Linnemann, J., Steinberger-Wilckens, R., 2007. Realistic costs of wind-hydrogen vehicle fuel production. Int. J. Hydrog. Energy 32, 1492–1499.

Martin, K.B., Grasman, S.E., 2009. An assessment of wind-hydrogen systems for light duty vehicles. Int. J. Hydrog. Energy 34, 6581–6588.

McCarthy, J.J., Canziani, O.F., Leary, N.A., Dokken, D.J., White, K.S., 2001. Climate Change 2001: Impacts, Adaptation and Vulnerability Contribution of Working Group II to the Third Assessment Report of the Intergovernmental Panelon Climate Change (IPCC). Cambridge University Press, Cambridge.

Meher, L.C., Sagar, D.V., Naik, S.N., 2006. Technical aspects of biodiesel production by transesterification a review. Renew. Sust. Energ. Rev. 10, 248.

Moriarty, P., Honnery, D., 2007. Intermittent renewable energy: the only future source of hydrogen? Int. J. Hydrog. Energy 32, 1616–1624.

Moriarty, P., Honnery, D., 2009. Hydrogen's role in an uncertain energy future. Int. J. Hydrog. Energy 34, 31–39.

Mueller-Langer, F., Tzimas, E., Kaltschmitta, M., Peteves, S., 2007. Techno-economic assessment of hydrogen production processes for the hydrogen economy for the short and medium term. Int. J. Hydrog. Energy 32, 3797–3810.

NAS, 2004. The Hydrogen Economy—Opportunities, Costs, Barrriers, and R&D Needs. National Research Council and National Academy of Engineering. National Academies Press, Washington, DC.

Ogden, J.M., 1999. Prospects for building a hydrogen energy infrastructure. Annu. Rev. Energy Environ. 24, 227e79.

Ogden, J.M., Yang, C., Johnson, N., Ni, J., Lin, Z., 2005. Technical and Economic Assessment of Transition Strategies toward Widespread Use of Hydrogen as an Energy Carrier. UCDITSRR-05-06.

Pearce, F., 2006. Special report: climate change. New Scientist.com News Service. See also, http://environment.newscientist.com/channel/earth/climate-change/dn9903S.

Pilavachi, P.A., Chatzipanagi, A.I., Spyropoulou, A.I., 2009. Evaluation of hydrogen production methods using the analytic hierarchy process. Int. J. Hydrog. Energy 34, 5294–5303.

Salgi, G., Donslund, B., Alberg Østergaard, P., 2008. Energy system analysis of utilizing hydrogen as an energy carrier for wind power in the transportation sector in Western Denmark. Util. Policy 16, 99–106.

Scott, D.S., 2004. Back from the future: to built strategies taking us to a hydrogen age. In: Sperling, D., Cannon, J.S. (Eds.), The Hydrogen Energy Transition: Moving Toward the Post Petroleum Age in Transportation. Elsevier, Amsterdam, pp. 21–32.

Shoko, E., et al., 2006. Hydrogen from coal: production and utilisation technologies. Int. J. Coal Geol. 65, 213–222.

Sperling, D., Cannon, J.S., 2004. Introduction and overview. In: Sperling, D., Cannon, J.S. (Eds.), The Hydrogen Energy Transition: Moving Toward the Post Petroleum Age in Transportation. Elsevier, Amsterdam, pp. 1–19.

Thomas, C.E., Kuhn, I.F., James, B.D., Lomax, F.D., Baum, G.N., 1998. Affordable hydrogen supply pathways for fuel cell vehicles. Int. J. Hydrog. Energy 23, 507–516.

Tsuchiya, H., 2008. Innovative renewable energy solutions for hydrogen vehicles. Int. J. Energy Res. 32, 427–435.

Turton, H., Barreto, L., 2006. Long-term security of energy supply and climate change. Energ Policy 34, 2232–2250.

Vidueira, J.M., Contreras, A., 2003. PV autonomous installation to produce hydrogen via electrolysis, and its use in FC buses. Int. J. Hydrog. Energy 28, 927–937.

Waegel, A., Byrne, J., Tobin, D., Haney, B., 2006. Hydrogen highways: lessons on the energy technology-policy interface. Bull. Sci. Technol. Soc. 26, 288–298.

Wang, Y.-W., Lin, C.-C., 2009. Locating road-vehicle refuelling stations. Transp. Res. E 45, 821–829.

Wietschel, M., Seydel, P., 2007. Economic impacts of hydrogen as an energy carrier in European countries. Int. J. Hydrog. Energy 32, 3201–3211.

Williams, R.H., 2004. Crude Oil, Climate Change, Coal, Cane and Cars. Princeton Environmental Institute, Princeton University.

Yildiz, B., Kazimi, M.S., 2006. Efficiency of hydrogen production systems using alternative nuclear energy technologies. Int. J. Hydrog. Energy 31, 77–92.

Chapter 3

Hydrogen Demand Side

1. HYDROGEN TODAY

In the scenario proposed by the International Energy Agency in November 2016, a 30% rise in global energy demand to 2040 means an increase in consumption for all modern fuels, but the global aggregates cover a multitude of diverse trends and significant switching between fuels. The World Energy Outlook, Executive Summary, asserts that, globally, renewable energy sees by far the fastest growth. Natural gas fares best among the fossil fuels, with consumption rising by 50%. It has been evaluated that growth in oil demand slows, but tops 103 million barrels per day (mb/d) by 2040 (World Energy Outlook (WEO), 2016).

In 2015, Renewable Energy Sources (RES) increased in power generation reaching 2.8% of global energy consumption with a growing of 0.8% in the last decade. Renewable energy used in power generation grew by 15.2%, slightly below the 10-year average growth of 15.9% but a record increment (+213 terawatt-hours), which was roughly equal to all of the increase in global power generation. Renewables accounted for 6.7% of global power generation (BP Statistical Review of World Energy, 2016).

Gross inland energy consumption in the EU-28 in 2014 was 3.6% lower than in 2013. Crude oil and petroleum products represent the most important energy source for the European economy, despite the long-term downward trend, while natural gas remains the second most important energy source. Contribution of RES is increasing; however, their contribution still did not surpass any of the fossil fuels (oil, gas, coal) nor the contribution of nuclear energy (Fig. 3.1).

Over the past decade (2004–14), fossil fuel and nuclear energy had negative trend in primary energy production. Production of petroleum products registered the biggest decrease (52.0%) while gas production fell by 42.9%. However, there was a positive trend in production of renewable energies over the same period, with a 73.1% increase (Eurostat (nrg_100a), 2016).

Navigant Research (http://www.navigantresearch.com/) forecasts that global demand for hydrogen from the power-to-power, power-to-transport, and power-to-gas sectors will reach 3.5 billion kg annually by 2030 (Market Data: Hydrogen Infrastructure, 2016).

From a normative view point, in 2014, the European Commission published the Directive 2014/94/EU (http://eur-lex.europa.eu/legal-content/EN/TXT/?qid=1481023032385&uri=CELEX:32014L0094) on the deployment of

Hydrogen Infrastructure for Energy Applications. https://doi.org/10.1016/B978-0-12-812036-1.00003-2
© 2018 Elsevier Inc. All rights reserved.

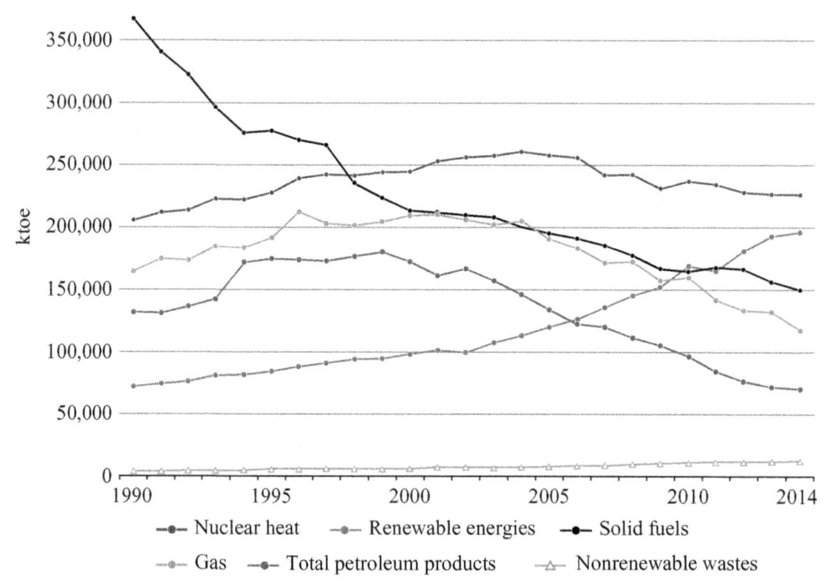

FIG. 3.1 Primary energy production. *(Source: http://ec.europa.eu/eurostat/statistics-explained/index.php/Energy_trends (accessed 06/12/2016).)*

alternative fuels infrastructure which reflected the Communication from the Commission of 24 January 2013 entitled "Clean Power for Transport: A European alternative fuels strategy." These documents identified hydrogen as alternative fuel to replace petrol in the long term because it also offers the technical potential to be used in simultaneous and combined dual-fuel technology systems.

Hydrogen integration in the energy system may have a potential impact directly on the key factors of the European energy policy. Improved system energy efficiency and emission reductions mean that hydrogen has the potential to reduce local and global emissions promoting environmental sustainability. In addition, hydrogen makes the economy more flexible and promotes a variety of primary energy sources which provide security in supply options; in particular, hydrogen can favor the use of higher share of RES in the energy sector. The development of market leading hydrogen technologies and the employment they bring with them means that the integration of hydrogen can contribute significantly to ongoing European economic competitiveness (European Commission EUR 23123, 2008).

2. EXISTING AND EMERGING MARKETS

Primary production of renewable energies is on a long-term increasing trend. Between 1990 and 2013 it increased by 170% (Energy, Transport and Environment Indicators, 2015).

Over the last decade, many EU partnerships of regional and municipal authorities have been created for emission reduction and clean technology integration, in 2008, with the support of the EU Commission, the European Regions and Municipalities Partnership for hydrogen and fuel cells, HyRaMP (www.hy-ramp.eu). HyRaMP was established to facilitate the development of the first markets for hydrogen and fuel cell applications by hosting and cofunding big demonstration projects and setting up joint procurement and local supply and maintenance chains. HyRaMP is now representing >30 regions and its secretariat is hosted by the EHA office in Brussels (Stationary Applications, n.d.).

In recent years, the ability of hydrogen to improve safety in transport, to balance the electricity grid and to allow an increased use of RES in transport and heat applications led to a positive market outlook for fuel cells and hydrogen technologies (FCH). In addition, demand in refining and chemical industries will probably increase due to a lower quality of crude oil, to the need for cleaner fuels and to the growing demand for fertilizers (Fig. 3.2) (JRC Science and Policy Reports Joint Research Centre Report, Technology Descriptions, 2013).

Market revenues from H2 for European mobility could reach several billions of euros by 2030. In addition to its use in Fuel Cell Electric Vehicles, hydrogen represents a suitable alternative propulsion fuel for other transport modes, with the exception of long-distance freight road, aviation, and sea shipping (EC Communication, 2013). The global demand for hydrogen fuel (FCEVs, buses, forklifts, uninterrupted power supply, scooters) is expected to reach over 0.4 Mt/year by 2020, reflecting a 2010–20 growth rate of 88% (SETIS chapter on Fuel Cells and Hydrogen: http://setis.ec.europa.eu/technologies/fuel-cells-and-hydrogen. Last accessed November 2017).

By 2050, H2 should be produced through carbon-free or carbon-lean processes. Hydrogen production by electrolysis is expected to considerably grow

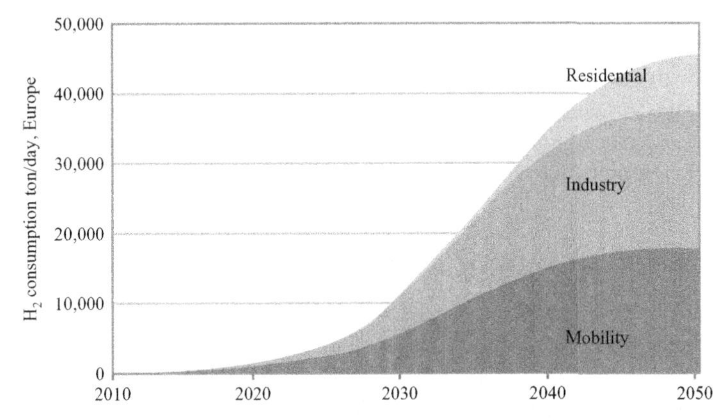

FIG. 3.2 Projection of the future hydrogen market in Europe. *(Adapted from Global CCS Institute Publication. Source: https://hub.globalccsinstitute.com/publications/2013-technology-map-european-strategic-energy-technology-plan-set-plan-technology-descriptions/133-market-and-industry-status-and-potential#fig_13.2. Last accessed January 2018.)*

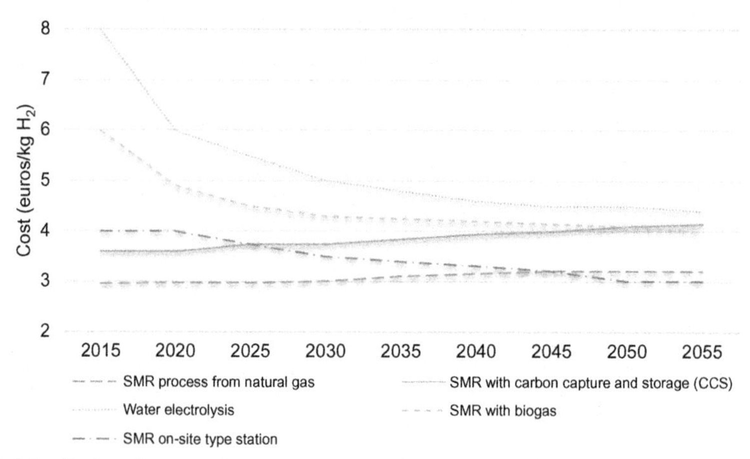

FIG. 3.3 Fuel cost by type of hydrogen production 2015–55 in EU-28. *(Adapted from Cantuarias-Villessuzanne, C., Weinberger, B., Roses, L., Vignes, A., Brignon, J.M., 2016. Social cost-benefit analysis of hydrogen mobility in Europe. Int. J. Hydrog. Energy 41 (42), 19304–19311.)*

due to its ability to contribute to grid stability obtained both in the supply and demand management. The latter is particularly attractive for small-scale electrolyzers sited at refueling stations and has the added advantage of not requiring a distribution infrastructure.

The feedstock cost and conversion technology, the plant size, the required purity level, and the method and distance for hydrogen deliveries determine the final cost of hydrogen for users.

Fig. 3.3 shows 2055 projected cost ranges according to the most promising hydrogen production technologies defined by the European project DEMCAMER: SMR process from natural gas, SMR with carbon capture and storage (CCS), water electrolysis, SMR with biogas, and SMR on-site type station (Market Data: Hydrogen Infrastructure, 2016).

However, the transition to hydrogen as a new energy carrier requires a series of investments and developments not only in energy supply and distribution, but also in policy measures that value the benefits to society, such as reduced CO_2 emissions, enhanced energy security through better grid integration of renewable energy sources, improved air quality, reduced noise, etc.

3. MODELING HYDROGEN DEMAND UNCERTAINTIES

The criterion related to the density of the hydrogen demand is a key factor since the concentration or not of the hydrogen demand will contribute to the choice of the hydrogen production and transportation mode. This concentration may depend on the future opened hydrogen market and on the density of population.

The nature and size of demand for hydrogen will be explored for stationary and transportation applications.

3.1 Transportation

The transitions in fuel supply infrastructure and alternative vehicle for a low carbon transport system may need to be encouraged and coordinated at the EU level in order to drive the market forward.

In Europe, the transport sector produces >20% of the total greenhouse gases emitted to the atmosphere in 2012 (Evangelisti et al., 2017). Alternative vehicles, i.e., electric vehicles (EVs) or fuel cell vehicles (FCVs), will be essential to guarantee carbon emission reduction targets.

Hydrogen utilization can cover different modalities in transportation including FCV as well as rail, maritime, and aviation applications, but in addition, required a deep hydrogen refueling infrastructure investigation.

It is clear that an appropriate regulatory framework and financial instruments have to be introduced, furthermore to the efforts computed in the last years, to support clean alternatives to the hydrogen market. Special interest is dedicated to road transport with the introduction of the hydrogen fuel cell vehicle (FCV) which offers the greatest potential for addressing EU climate change and energy security objectives. FCVs represent a clean alternative for the different kinds of road traffic, whether urban, intercity, or long distance; anyway, FCVs are well suited for the passenger cars segment, which are responsible for about 12% of all CO_2 emissions in the EU.

A fuel cell is a device that produces electricity by an electrochemical reaction of hydrogen stored on the vehicle with oxygen from the air. Fuel cells are usually classified according to the type of electrolyte used in the cells such as PEMFC (proton exchange membrane fuel cells), AFC (alkaline fuel cell), MCFC (molten carbonate fuel cell), SOFC (solid oxide fuel cell), and PAFC (phosphoric acid fuel cell). The PEMFCs are considered the main promising technology for the FCVs due to high efficiency and environment friendly. Several automakers have announced plans to begin producing FCEVs for sale in the 2015–18 time period. Hyundai started FCEV production in 2014, and Toyota announced plans to begin producing FCEVs in the 2014–15 time period. Mercedes, Ford, and Nissan announced a joint FCEV development plan in 2013 with a goal of starting commercial production of FCEVs by 2017 (http://eur-lex.europa.eu/legal-content/EN/TXT/?qid=1481023032385&uri=CELEX:32014L0094).

The hydrogen FC system proposed by Mercedes comprises the fuel cell stack which converts the hydrogen and oxygen's air into electricity, water, and heat; the air system module which supplies the necessary air at the right pressure; the hydrogen module which supplies the right hydrogen quantity and the power distribution module which connects the stack bus bars and the high voltage components.

Demand side motivates FCV penetration market. The availability and the number of hydrogen refueling station, the performances, the distance traveled by FCV, and the costs for final user determine the changes in total hydrogen demand. Other important parameters can influence the transaction to hydrogen demand in respect to the conventional internal combustion engine vehicles (ICEV). Table 3.1 resumes the key "demand parameters" for a light-duty vehicle.

The possibility for users to recharge their vehicles with different options becomes essential in the prospective to replace ICEV. In October 2014, the European Parliament emanated the Directive 2014/94/EU. The aim of the Directive is to build up refueling points for alternative fuels (electricity, liquid natural gas, compressed natural gas, hydrogen) in the EU, with common standards for their design and use. Different methodology exists to identify the optimal location of refueling points for FCV automotive market in a municipality, both in terms of location and size. The directive requires Member States to set targets for recharging points accessible to the public, to be built by 2020, to ensure that electric vehicles can circulate at least in urban and suburban agglomerations. Most FVC cars available in 2014 have, from a fully charged battery, a driving range of approximately 600 km per fill-up depending on various factors such as weather conditions, traffic congestion, and road types (Hyundai, 2013). Besides, on-board vehicle storage of hydrogen still needs further development to increase both volumetric and gravimetric storage capacity. Different options exist to transport hydrogen from the central production site to ECPs. These include pipelines and trucking compressed gas or liquid hydrogen. A number of studies have however shown that distributed hydrogen production directly at refueling stations by steam reforming will supply hydrogen at the lowest cost by 2020.

TABLE 3.1 Demand parameters for a FCV vehicle diffusion.

Demand category	Customers	Regulatory
Parameter	Initial cost	Emissions of pollutants
	Operational and maintenance costs	Fuel efficiency
	Quality	Greenhouse gas emissions
	Refueling convenience	Safety
	Passenger/cargo space	
	Vehicle performances	
	Safety	

Adapted from The Hydrogen Economy: Opportunities, Costs, Barriers, and R&D Needs. National Academic Press. https://doi.org/10.17226/10922.

As a matter of fact, the daily behavior of people plays an important role in the analysis of how FCVs have to be charged and discharged (Hess et al., 2012). In fact, Zhao et al. (2011) demonstrated that if travelers, charging only at home, use FCV it may not fulfill the energy demand of a daily trip. Thus, additional charging infrastructures are necessary to be established on the territory to satisfy FCV users.

In a near-term scenario, it is presumable that infrastructure providers will add hydrogen fuel to an existing fueling station by steam methane reforming (SMR) process from natural gas if their forecast fuel revenue could deliver a desired revenue (Harrison and Thiel, 2017). Since many of the existing petrol service stations are already subject to safety criteria in terms of their locations, it is reasonable to assume that, in the years to come, some of them will be set up to recharge FCVs.

In literature, it is ascertained that policies to introduce sufficient public charging infrastructure are necessary to encourage the introduction of FCVs (Bakker and Jacob Trip, 2013).

Authors in (Harrison and Thiel, 2017) tried to identify what impact policy options may have on long-term FCVs penetration. They analyze numerous policy scenarios, recognizing that single emobility policies should not be considered in isolation as the interaction between multiple policies is highly relevant. The proposed approach seeks to understand how specifically supporting the infrastructural system may characterize uptake within the wider policy environment. For the wider policy environment, they considered supply (e.g., fleet emission regulation) and demand stimulating policies (e.g., purchase incentives). To do this, the research employed an extensive system dynamics model of the EU automobile market, which reflects the relevant market agents of users, manufacturers, infrastructure providers, and authorities. The number of desired fueling stations to invest in is therefore determined by using a similar calculation as in (Harrison and Thiel, 2017).

$$\text{Desidered H2 refilling station} = \frac{\text{Forecast revenue} - \text{Current revenue}}{\text{Running costs} + \text{Installation cost}x(1 + \text{Desidered ROI})}$$

where

- Forecast fuel demand is based on a 3-year forecast stock of the relevant power train and annual fuel consumption;
- Current and forecast revenue are based on a fuel cost and calibrated fuel margin;
- Investment costs reduce over time depending on the current installed base of hydrogen fueling stations;
- ROI = desired return on investment.

The EU Fuel Cell and Hydrogen Joint Undertaking Multi-Annual Work Plan (MAWP) targets an overall cost of H2 at the refueling station (including production, delivery, compression, storage, and dispensing, but exclusive of taxes) of EUR 5–9/kg in 2020 and EUR 4.5–7.0/kg in 2023, from a 2012 status of EUR 5.0 to over 13/kg (Fuel Cells and Hydrogen Joint Undertaking (FCH JU), 2014).

The number of H2 filling stations increased significantly in the world and, in fact, also in Europe, 19 new hydrogen fuel stations opened in 2015. Plans were also announced in the next years for 64 more hydrogen stations with 34 of those to be located in Germany.

The website Netinform.net (last accessed November 2017) provides a worldwide map which displays the existing hydrogen filling stations (http://www.netinform.net/H2/H2Stations/H2Stations.aspx?Continent=EU&StationID=-1).

3.2 Stationary Application

The use of hydrogen in stationary applications represents an important role in the development of hydrogen economy. Hydrogen production for energy storage and grid balancing from renewable electricity is one of the main priority in the stationary applications, including large "green" hydrogen production, storage, and reelectrification systems.

The facilities in which the fuel cells operate at a fixed location for primary power, backup power, or combined heating and cooling (CHP) represent stationary power application. The stationary sector includes both large scale (200 kW and higher) and small scale (up to 200 kW) and a wide range of markets including retail, data centers, residential, telecommunications, and many more.

Fuel cells convert the energy carrier hydrogen into electric power and heat. The cogeneration of electric power and heat makes the best possible use of the originally used primary energy carrier. Such cogeneration fuel cell power stations can be realized in different construction sizes. Beside small decentralized power stations with a power range between 200 kW and some megawatt, especially small systems are an interesting option. Within the power range of common home heating these systems do not only deliver heat but also electric power to feed into the power grid. Millions of such residential fuel cells can be seen as a large power plant, a power plant which can, for example, also serve as backup unit in a wider supply scenario mainly based on renewable energies. Today residential fuel cell systems for the homes work with natural gas and need a reformer to produce the hydrogen needed. In Germany many manufacturers of heating systems are working on systems like that.

The main challenge is to increase lifetime and performances for fuel cell systems for CHP or for power only, covering the technical requirements necessary to reduce costs. Hydrogen storage, handling, and distribution to allow storage of hydrogen at central production plant and distribution to the customer base is one of the main topics for hydrogen applications.

If technologies will reach adequate commercial level, FCs and turbines using hydrogen will become an important opportunity for hydrogen produced from sources other than natural gas in areas where pipelines are available. Anyway, this assumption implies consistent infrastructure investments to stimulate this market.

The possibility to use hydrogen to generate electricity on-site or transport and use it for industrial processes could represent another demand to stimulate a market for hydrogen production.

The 2017 Fuel Cell Technologies Market Report shows that splitting the fuel cell shipments by application, transport option increases sensibly. The total stationary units shipped increased, from about 52,000 to roughly 56,000 units, about 50,000 of which were in Japan. All stationary units together contributed 214 MW of power, very close to 2016 figures.

In the stationary sector, new generation fuel cell systems produce greater power and often customers are deploying larger systems. Several facilities in the United States and in Korea are installing multimegawatt, large-scale fuel cell power parks. U.S. fuel cell manufacturers are exporting these fuel cell systems to Korea, accounting for much of the 20% growth in fuel cell shipments (Fig. 3.4).

For small application, in Europe, Viessmann joined the ene.field residential fuel cell CHP project, which aims to deploy up to 1000 fuel cells in 11 European countries. Ene.field is the largest European demonstration of the latest smart energy solution for private homes, fuel cell micro-CHP. As of October 2015, >350 units have been installed in seven countries—Italy, Germany, Denmark, France, Australia, Luxembourg, and Switzerland (Fuel Cell Technologies Market Report, 2015). In this context, also SOFT-PACT project will provide

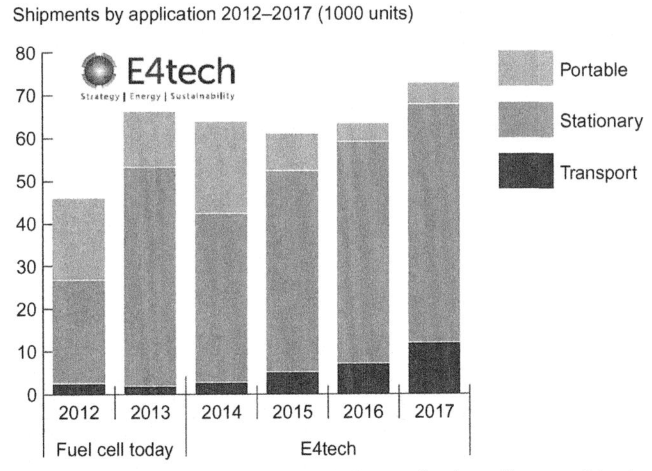

FIG. 3.4 Fuel cell system shipper worldwide by application. *(Source: E4tech. Fuel Cell Technologies Market Report 2017. http://www.FuelCellIndustryReview.com.)*

Examples of commercially available hydrogen generation systems 2015			
Manufacturer	**Product**	**Type**	**Hydrogen production**
Acta S.p.A. Italy	EL Series	Alkaline solid polymeric electrolytic process	0.54–2.2 kg/day
HyGear Netherlands	HyGEN Series	Reformer	10.8–225 kg/day
ITM Power U.K.	HPac Series	PEM electrolysis	2.5–5 kg/day
ITM Power U.K.	HGAS	PEM electrolysis	25–462 kg/day
Linde Germany	HYDROPRIME	Reformer	713–2160 kg/day
McPhy France	Baby McPhy	Alkaline electrolysis	0.86 kg/day
McPhy France	McLyzer	Alkaline electrolysis	2.2–43 kg/day
McPhy France	Large H2 production units	Alkaline electrolysis	216–865 kg/day
Siemens Germany	SILYZER Series	PEM electrolysis	48–5400 kg/day

FIG. 3.5 Examples of commercially available hydrogen generation systems. *(Source: Fuel Cell Technologies Market Report. https://www.energy.gov/sites/prod/files/2016/10/f33/fcto_2015_market_report.pdf.)*

65 fuel cell systems with electrical efficiency higher than 42% over lifetime (total efficiency higher than 78%) and 25% cost reduction.

In Europe, the four European Projects HELTSTACK, NELLHI, INNO-SOFC, and DEMOSOFC dedicated special attention to SOFC technology within an all-European supply chain for next-generation, clean heat and power.

In the Commercial Market Segment (5–400 kW), the SOFCOM project provides proof-of-concept poly-generation SOFC systems fed by biogenous primary fuels (biogas and bio-syngas, locally produced), modular concept, cost driver identified.

For the Industrial Market Segment (>0.3 MW) the POWER-UP project will provide the first module of 40 kW (out of 240 kW) with 61% electrical efficiency (started October 2015). The ClearGenDemo project considers to install near Bordeaux (FR) 1 MW PEM on by-product H2 from clori-alkali plant. Finally, DEMCOPEM-2 MW project will provide a 2-MW PEM (European technology) to be demonstrated in China (Atanasiu, 2015).

Fig. 3.5 provides examples of commercially available hydrogen generation systems in Europe. These systems produce hydrogen using either electrolysis (running an electrical current through water to produce hydrogen) or by reforming hydrocarbon or bio-based fuels (a process in which methane from the source fuel is heated, with steam, to produce hydrogen).

4. CONCLUSION

To obtain a significant transition toward hydrogen in the future energy systems, the hydrogen production, distribution, storage, and use must be greatly improved in respect with their current performance, reliability, and cost.

The timescale and evolution of such a transition is the focus of many "roadmaps" emanating from the United States, Japan, Canada, and European Commission. The main highlights and scenarios of the major European technology platform are summarized in Fig. 3.6.

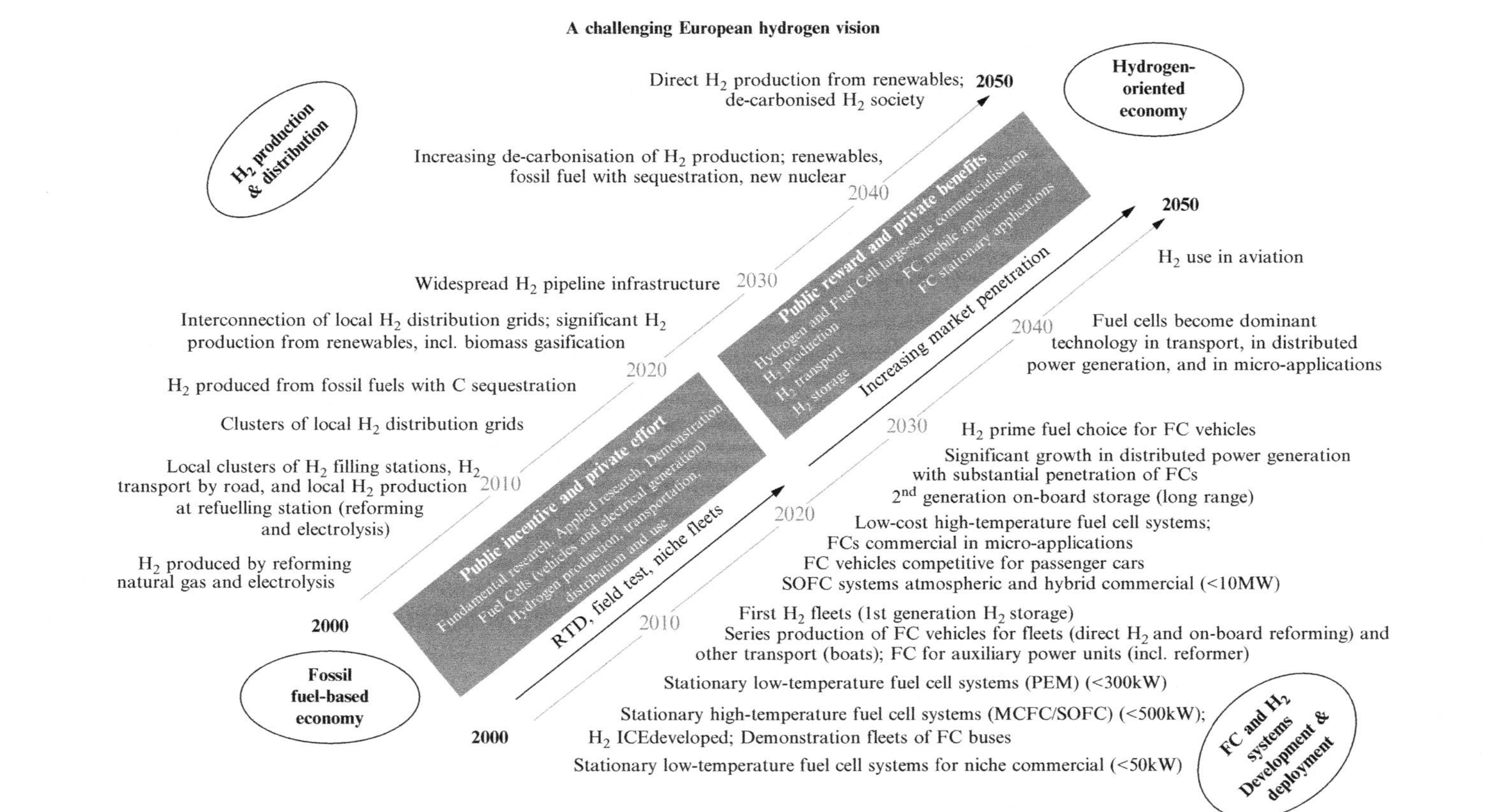

FIG. 3.6 European hydrogen roadmap. *(Adapted from Edwards, P.P., Kuznetsov, V.L., David, W.I.F., Brandon, N.P., 2008. Hydrogen and fuel cells: towards a sustainable energy future. Energy Policy 36 (12), 4356–4362. ISSN 0301-4215. https://doi.org/10.1016/j.enpol.2008.09.036.)*

Some of the key scientific and technical challenges for the development of hydrogen economy are summarized as follows (Edwards et al., 2008):

- lowering the cost of hydrogen production to a level comparable to the energy cost of petrol,
- development of a CO_2-free route for mass production of sustainable hydrogen at a competitive cost,
- development of a safe and efficient national infrastructure for hydrogen delivery and distribution,
- introduction of praticable hydrogen storage systems for both vehicular and stationary applications and
- dramatic reduction in costs and significant enhancement in the durability of fuel cell systems.

Besides, a long-term and sustained research and development policies are required to address the remaining high-risk technological barriers for a powerful hydrogen economy penetration. This includes, in particular, the cost reduction, new engineering concepts, and alternative approaches in different application areas (portable, stationary, and transport). Even if the market could be ready, the technology is not enough and the hydrogen transition must be accompanied by policies and incentives that involved public and private stakeholders. Because of the interdependence of technological progress and such policy measures in order, the activities needed to ensure commercial sustainability can only be attained through public–private partnerships.

From regulatory incentives viewpoint, the NEW-IG, Hydrogen Europe, the leading European industry association representing almost 100 companies in the fuel cells and hydrogen industry, published that it is working with policy-makers to realize the following milestones (Harrison and Thiel, 2017):

Hydrogen production, handling & distribution

- Ensure favorable tax regime for decarbonized hydrogen to incentivize market development.
- Recognize sustainable hydrogen coming from renewable energy or produced with carbon capture and storage (CCS).

Grid flexibility

- Acknowledgment of hydrogen production from the grid or storage as a value-adding component for the energy infrastructure.
- Introduce regulation to introduce hydrogen into the natural gas network.

Stationary fuels cells

- Remove the uncertainty according to the classification of gas-based stationary fuel cells under the Energy Labelling Directive.
- Adapt the Network Code on Requirements for Grid Connection Applicable to All Generators to account for the particularities of small-scale emerging technologies like fuel cell micro-CHP.

Transport incl. power to fuel

- Increase support for renewable, decarbonized hydrogen as clean alternative fuel for transport (Renewable Energy Directive—RED).
- Complete the list of hydrogen production strategies in the reporting methodology of the Fuel Quality Directive—FQD).
- Develop minimum mandatory targets at Member State level for Hydrogen Refueling Stations Deployment (clean power for transport package—CPTP).

REFERENCES

Atanasiu, M., 2015. Introduction to portfolio of Energy, Programme, Review Days 2015. Available at http://www.fch.europa.eu/sites/default/files/Mirela-Energy_portfolio%20%28ID%202848664%29.pdf (last accessed November 2017).

Bakker, S., Jacob Trip, J., 2013. Policy options to support the adoption of electric vehicles in the urban environment. Transp. Res. Part D: Transp. Environ. 25, 18–23.

BP Statistical Review of World Energy, June 2016. bp.com/statisticalreview.

EC Communication, 2013. Clean Power for Transport: A European Alternative Fuels Strategy.

Edwards, P.P., Kuznetsov, V.L., David, W.I.F., Brandon, N.P., 2008. Hydrogen and fuel cells: towards a sustainable energy future. Energy Policy 36 (12), 4356–4362. ISSN 0301-4215, https://doi.org/10.1016/j.enpol.2008.09.036.

Energy, Transport and Environment Indicators. http://ec.europa.eu/eurostat/documents/3217494/7052812/KS-DK-15-001-EN-N.pdf/eb9dc93d-8abe-4049-a901-1c7958005f5b.

European Commission EUR 23123, 2008. The European Hydrogen Roadmap. Office for Official Publications of the European Communities.

Eurostat (nrg_100a) 2016 (accessed December 2016).

Evangelisti, S., Tagliaferri, C., Brett, D.J.L., Lettieri, P., 2017. Life cycle assessment of a polymer electrolyte membrane fuel cell system for passenger vehicles. J. Clean. Prod. 142 (Part 4), 4339–4355. ISSN 0959-6526. https://doi.org/10.1016/j.jclepro.2016.11.159.

Fuel Cell Technologies Market Report. https://www.energy.gov/sites/prod/files/2016/10/f33/fcto_2015_market_report.pdf. Accessed 1 October 2017.

Fuel Cells and Hydrogen Joint Undertaking (FCH JU), 2014. Multi-Annual Work Plan 2014–2020. http://www.fch.europa.eu/sites/default/files/documents/FCH2%20JU%20-%20Multi%20Annual%20Work%20Plan%20-%20MAWP_en_0.pdf.

Harrison, G., Thiel, C., 2017. An exploratory policy analysis of electric vehicle sales competition and sensitivity to infrastructure in Europe. Technol. Forecast. Soc. Chang. 114, 165–178. ISSN 0040-1625.

Hess, A., Malandrino, F., Reinhardt, M.B., Casetti, C., Hummel, K.A., Barceló-Ordinas, J.M., 2012. Optimal deployment of charging stations for electric vehicular networks. In: Proceedings of the First Workshop on Urban Networking (UrbaNe'12), Nice, France. ACM, New York, NY, pp. 1–6. https://doi.org/10.1145/2413236.2413238.

Hyundai, 2013. Hyundai ix35 fuel cell to demonstrate real-world benefits to EU decision-makers. http://www.hyundai.co.uk/about-us/environment/hydrogen-fuel-cell#technology.

JRC Science and Policy Reports Joint Research Centre Report, Technology Descriptions, 2013. Technology Map of the European Strategic Energy Technology Plan. Report EUR 26345 EN.

Market Data: Hydrogen Infrastructure. https://www.navigantresearch.com. Accessed December 2016.

Stationary Applications. http://www.h2euro.org/wp-content/uploads/2009/10/EHA_Contribution_EU_Energy_2011_2020_final.pdf.

The Hydrogen Economy: Opportunities, Costs, Barriers, and R&D Needs. National Academic Press. https://doi.org/10.17226/10922.

World Energy Outlook (WEO), 2016. International Energy Agency (IEA). https://www.iea.org/newsroom/news/2016/november/world-energy-outlook-2016.html.

Zhao, L., Awater, P., Schäfer, A., Breuer, C., Moser, A., 2011. Scenario-based evaluation on the impacts of electric vehicle on the municipal energy supply systems. In: Power and Energy Society General Meeting, 2011 IEEE, July. pp. 1–8.

FURTHER READING

Cantuarias-Villessuzanne, C., Weinberger, B., Roses, L., Vignes, A., Brignon, J.M., 2016. Social cost-benefit analysis of hydrogen mobility in Europe. Int. J. Hydrog. Energy 41 (42), 19304–19311.

Fuel Cells Network Webinar, 2016. European-Wide Deployment of Residential Fuel Cell Micro-CHP. Fuel Cells for Stationary Power Applications. http://www.fuelcellnetwork.eu/wp-content/uploads/Ene-field-Presentation-2016.pdf.

The ultimate guide to fuel cells and hydrogen technology, Hydrogen Europe. www.hydrogeneurope.eu (http://hydrogeneurope.eu/wp-content/uploads/2016/02/FCH_Brochure_V2SP.pdf, 01 November 2017). See, http://hydrogeneurope.eu/wp-content/uploads/2016/02/FCH_Brochure_V2SP.pdf. Accessed 1 November 2017.

Chapter 4

Hydrogen Storage and Distribution: Implementation Scenarios

1. INTRODUCTION

In Europe, different regulations, standards, and codes exist to guarantee safe and reliable design and operation of a product, facility, and equipment, which involve hydrogen.

European Community Directives represent in Europe legal regulations for manufacturing, operation, and maintenance of hydrogen installations. Codes and standards, which are not legal documents, have the function of guidelines, often realized by interested stakeholders, to support safety and common understanding. European and USA codes are often used differently (Roads2HyCom, 2008).

CEN (European Standardisation Committee) and CENELEC (European Committee for Electrochemical Standardisation) represent the European counterparts to the International Standardisation Organisations ISO (International Standardisation Committee) and IEC (International Electrical Committee). They have the main role to adopt international standards and to harmonize these with a range of the European directives.

Fig. 4.1 shows an overview of the international work on standardization of hydrogen related to equipment and technologies (European Commission, 2006).

Recently, Europe entered into force the European Directive 2014/68/EU PED (Pressure Equipment Directive) on 20 July 2016. The Pressure Equipment Directive covers a very large range of products such as vessels, pressurized storage containers, heat exchangers, steam generators, boilers, industrial piping, safety devices, and pressure accessories. Besides, two ATEX directives (Directives 94/9/EC-ATEX 95 & 1999/92/EC-ATEX 137) regulate equipment intended for use in potentially explosive atmospheres (ATEX 95) and H&S protection measures for workers potentially at risk from explosive atmospheres (ATEX 137).

The main organizations that provide information about standard safety measures are the International Standards Organisation (ISO/TC 197) (Technical Committee 197 "Hydrogen Technologies"), the International Electrotechnical

Hydrogen Infrastructure for Energy Applications. https://doi.org/10.1016/B978-0-12-812036-1.00004-4
© 2018 Elsevier Inc. All rights reserved.

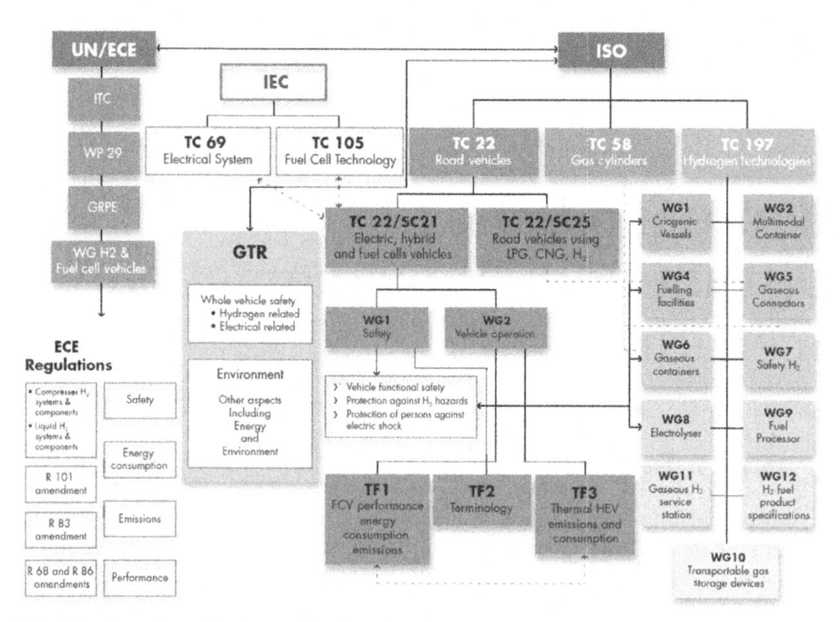

FIG. 4.1 Overview of the international work on standardization of hydrogen related to equipment and technologies (European Commission, 2006).

Commission (IEC) (Technical Committee 105 "Fuel Cell Technologies"), the European Industrial Gases Association (EIGA), the Compressed Gas Association (CGA), and the National Fire Protection Association (NFPA).

The ISO TC197 "Hydrogen Technologies" covers different topics relating to Hydrogen Safety: Basic Properties; Materials/embrittlement; Combustion and Cryogenic Issues.

United Nations Economic Commission for Europe Global Technical Regulation (GTR) Number 13 (*Global Technical Regulation on Hydrogen and Fuel Cell Vehicles* (Global Technical Regulation, n.d.)) is a document which contains the safety requirements for freight hydrogen vehicles, and in particular fuel cell electric vehicles (FCEV). GTR Number 13, already formally adopted, represents a reference for national regulation regarding FCEV safety in North America (led by the United States), Japan, Korea, and the European Union.

2. HYDROGEN STORAGE REQUIREMENTS

In the current state of the art in hydrogen storage, any technology satisfies completely manufacturers and end users requirements, and a large number of obstacles have to be overcome.

Hydrogen contains a lot of energy per unit mass while the content of energy per unit of volume is quite low. This implies a potential problem in terms of storing large amounts of hydrogen.

The gravimetric density (GD) and volumetric density (VD) of the storage systems represent the techniques to evaluate hydrogen storage performance. The GD is the weight percentage of hydrogen relative to the total weight of the system (hydrogen + storage medium). The VD is the mass of hydrogen stored per unit volume of the system (Durbin et al., 2016).

Hydrogen has a very low density as both a gas and a liquid. Its density is 0.089 kg/m^3 in normal temperature (20°C) and pressure (1 atm), i.e., 7% of the density of air, 70.8 g/L as liquid (at −253°C), i.e., 7% of the density of water, and 70.6 g/L as solid (−262°C). For comparison, the density of gasoline has a density of 4.4 kg/m^3 in gaseous state and 700 kg/m^3 at liquid state (Prabhukhot et al., 2016) .

Another important property of hydrogen is the auto-ignition temperature, which is the lowest temperature at which it spontaneously ignites in normal atmosphere without an external source of ignition, such as a flame or spark. This temperature is required to supply the activation energy needed for combustion. This temperature for hydrogen is 585°C, higher than any other conventional fuel such as methane (540°C) and gasoline (230–480°C) (U.S. Department of Energy, n.d.). Hydrogen's flammability range (between 4% and 75% in air) is very large compared to other fuels (Fig. 4.2). Under the optimal combustion condition (a 29% hydrogen-to-air volume ratio), the minimum energy required for ignition in air is 0.02 MJ, almost an order of magnitude lower than other conventional fuels, i.e., 0.24 MJ for gasoline and 0.29 MJ for natural gas (Das et al., 2000). However, at low concentrations of hydrogen in air, the energy required to generate combustion is similar to that of other fuels.

Liquid hydrogen has different characteristics and potential hazards than gaseous hydrogen, so different control measures have to be adopted to guarantee safety. As a liquid, hydrogen is stored at −423°F, a temperature that can cause cryogenic burns or lung damage. Due to its very low viscosity hydrogen leaks at high speed in defective tanks. Detection sensors and tracing and trucking monitoring systems are critical when dealing with a potential liquid hydrogen leak or spill.

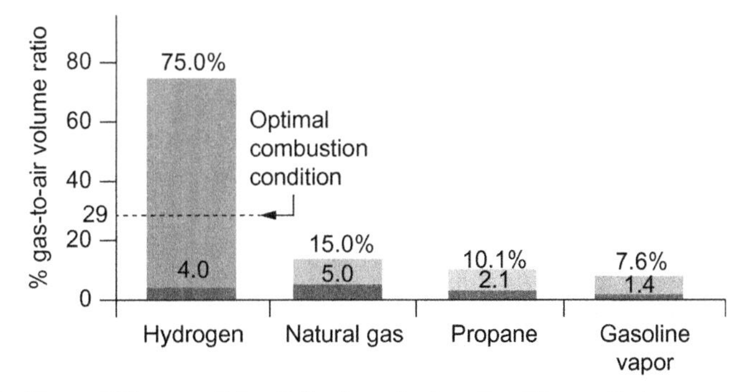

FIG. 4.2 Flammability range (https://h2tools.org/bestpractices/h2properties).

If an accident occurs involving hydrogen, the main potential ignition sources that can ignite a hydrogen leak may be the following:

- electrical (e.g., static electricity, electric charge from operating equipment)
- mechanical (e.g., impact, friction, metal fracture)
- thermal (e.g., open flame, high-velocity jet heating, hot surfaces, vehicle exhaust)

According to NFPA 55, both compressed gaseous hydrogen storage vessels and liquid hydrogen storage vessels must be located at least 50 feet from combustible materials.

The U.S. Department of Energy created a H_2 accident database in order to facilitate the sharing of lessons learned and other relevant data coming from actual experiences using and working with hydrogen (https://h2tools.org/lessons). The safety event records have been contributed by a variety of global sources, including industrial, government, and academic facilities. The database also serves as a voluntary reporting tool for capturing records of events involving either hydrogen or hydrogen-related technologies. Storage and handling of compressed hydrogen gas and cryogenic liquid hydrogen present potential health and safety hazards. Use of proper storage and handling techniques is essential to maintaining a safe work environment.

Different safety measures can be applied to mitigate, control, or prevent the accident, which involves hydrogen, or to protect people or equipment from the consequences of a given event.

In general, some examples of protection measures to limit amount of flammables and dispersion process are (HySafeVersion 1.0, 2006) as follows:

- Confine leak exposed zone both by solid casing and by soft barriers (polyethylene sheets). This may limit flammable cloud size, by physically limiting the cloud or reducing the momentum of a jet release.
- Decrease confinement near leak exposed area to allow buoyancy-driven dispersion transporting hydrogen away.
- Natural, forced and emergency ventilation to remove hydrogen
- Removal of ignition sources to limit explosion frequency.
- Igniters (or continuous burners) to ensure that gas clouds are ignited before they grow too large to limit consequences.
- Catalytic recombiners to remove unwanted hydrogen.
- Reducing the reactivity, inert gas dilution after release but prior to ignition.
- Fine water mist dilution to decrease flammability or sprinklers to improve mixing/dilution
- Rapid injection of dense hydrocarbon gas (e.g., butane) with much lower reactivity than hydrogen.
- Detection, activate shut-down (ESD), pressure relief, and safety measures, move people to safe place. Fire, limiting fire loads and consequences:
- Proper design against heat loads

- Passive fire protection to shield equipment and increase time before escalation
- Sprinkler systems and water deluge to cool equipment and control flames
- Inert gas systems or fine water mist to dilute oxygen and reduce heat generation.
- Avoid feeding oxygen into fire by proper confinement, limit ventilation.

3. HYDROGEN STORAGE TECHNOLOGIES

A feasible hydrogen economy has to be supported by the development of an efficient and economical method of storing and transporting hydrogen.

Two major technologies exist for hydrogen storage: storage in containers and in materials. Storage in containers includes the compressed hydrogen storage and liquid hydrogen storage approaches. In the compressed hydrogen storage approach, hydrogen is saved in the form of gas, with high pressure. In contrast, in the liquid hydrogen storage approach, hydrogen is saved in liquid form. On the other hand, storing hydrogen in materials involves the use of metal hydride, sorption materials, and chemical hydride.

3.1 Compressed Gaseous Hydrogen

Compressed hydrogen is considered to be a solution for hydrogen storage on automotive systems due to the relative simplicity of gaseous hydrogen, rapid refueling capability, excellent dormancy characteristics, and low infrastructure impact. One downside of the methods is a significant energy penalty—up to 20% of the energy content of hydrogen is required to compress the gas.

At the present, four types of pressure vessels exist to store hydrogen (Barthelemy et al., 2017; U.S. Drive, 2013).

Type	Technological characteristics	
Type I	Pressure limited to 50 MPa	All-metal cylinder
Type II	Pressure not limited	Load-bearing metal liner hoop wrapped with resin-impregnated continuous filament
Type III	For $P \leq 45$ MPa	Nonload-bearing metal liner axial and hoop wrapped with resin-impregnated continuous filament
Type IV	For $P \leq 100$ MPa	Nonload-bearing, nonmetal liner axial and hoop wrapped with resin-impregnated continuous filament

The shape of the vessels is generally cylindrical but they can also be polymorph or toroid.

Storing 1 kg of hydrogen at 100 kPa and 25°C requires a tank of volume 12.3 m^3. Compressing hydrogen to 350 bars decreases the required storage volume by 99.6%. Further pressure increases lower the required storage volume, but increase the compression work input and safety concerns (Hosseini et al., 2012). A key enabler factor to use this route of storage is the public's awareness on safety issues associated with high pressure hydrogen tanks.

Compressed hydrogen can be stored in closed tank systems at volumetric densities of around 20–50 kg/m^3 and gravimetric densities (kg H_2/kg of the tank) of around 5%–10% (Hosseini et al., 2012).

In the case of automotive systems, it was internationally agreed upon a standard pressure for the storage of gaseous hydrogen in the car of 70 MPa (700 bar). The hydrogen quality for automotive applications is regulated in ISO14687-2 (Hosseini et al., 2012). ISO 12619-1:2014 specifies general requirements and definitions of compressed gaseous hydrogen (CGH2) and hydrogen/natural gas blends fuel system components, intended for use on the types of motor vehicles defined in ISO 3833 (https://www.iso.org/standard/51569.html). It also provides general principles and specific requirements for design, instructions, and markings.

The hydrogen tanks at 700 atm have been designed with four-layers structure in order to guarantee driving safety. The aluminum-alloy tank is lined internally with plastic lining and wrapped externally in a protective layer of carbon fiber-reinforced plastics, with one more shock-absorbing protective layer of fiber glass material added outside that protective layer (He et al., 2016). The volumetric density of n-H_2 (normal hydrogen) as a function of the pressure for three different temperatures can be displayed as in Fig. 4.3 (Barbier, 2010).

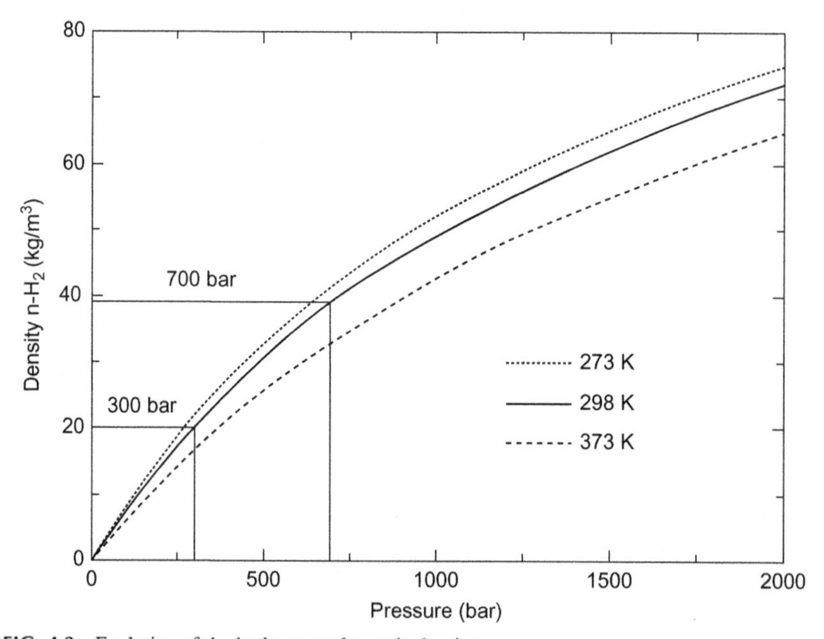

FIG. 4.3 Evolution of the hydrogen volumetric density.

Stationary high pressure gaseous hydrogen (HPGH2) storage vessel is mostly used to store hydrogen in hydrogen refueling stations for automotive market. In this case, large-scale, low-cost hydrogen storage is required. Two different solutions exist for stationary HPGH2 storage vessels, namely, seamless hydrogen storage vessel and multifunctional layered stationary hydrogen storage vessel (Zheng et al., 2012). Seamless pressure vessels made from high strength steel are usually small in volume, more susceptible to hydrogen embrittlement, and difficult for online safety monitoring. On the contrary, a multilayered steel high-pressure hydrogen storage vessel is flexible in design, convenient in fabrication, safe in use, wide in feasibility, and easy in online safety monitoring (Zheng et al., 2012).

Recently, the technical specification ISO/TS 19880-1:2016 (*Gaseous hydrogen—Fuelling stations*—Part 1: *General requirements)* has been defined as guideline on safety and performance for hydrogen fueling stations (https://www.iso.org/news/2016/07/Ref2104.html) and to contribute to their worldwide deployment.

3.2 Liquid Hydrogen

Another storage type is the one based on storing hydrogen in liquid form. Hydrogen can be stored in liquid form at extremely low temperature for both stationary and onboard vehicle applications. To be liquefied, up to 40% of energy content of hydrogen is required.

Hydrogen liquefaction and use of liquid hydrogen is usually adopted only when high storage density is required, for example, in aerospace applications. Some prototype hydrogen-powered automobiles as well as commercially available automobiles also use specially developed liquid hydrogen tanks.

Liquid hydrogen is stored in cryogenic tanks at $-253°C$ at ambient pressure. The advantage for the cryogenic liquefied hydrogen (LH2) is that it has a considerably higher energy content per unit of volume than compressed hydrogen gas, and therefore it needs less storage space (see diagram (Linde Group, n.d.)) (Fig. 4.4).

The volumetric density of liquid hydrogen is $70.8\,kg\,m^3$ and higher than that of solid hydrogen ($70.6\,kg\,m^3$). The advantage of liquid hydrogen storage are the energy-efficient liquefaction process and the thermal insulation of the cryogenic storage vessel in order to reduce the boil-off of hydrogen (Zuttel, 2004).

3.3 Solid Material

As alternative to the traditional mechanical storage (storage in a tank of compressed gas or liquid hydrogen) hydrogen can be stored:

- by physisorbtion (*physical adsorption*), in porous and solid materials which include graphene and other carbon structures, metals and metallic

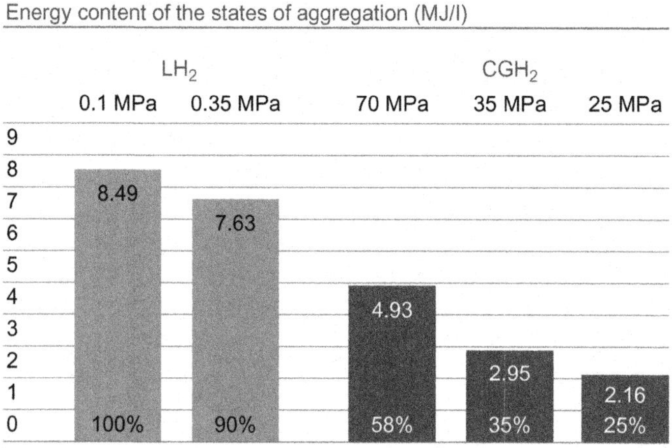

FIG. 4.4 Energy content (MJ/l) of the various states of aggregation of liquid hydrogen (LH2) and compressed hydrogen gas (CGH2) at various pressures.

nanocrystals and composites as metal hydrides, metal-organic frameworks (MOF), zeolites;

- in solid or liquid material of chemically bound hydrogen that is released on decomposition which comprises light metal hydrides (e.g., alkaline hydrides, alanates, alane), borohydrides, amines and imides, amino borane.

Certain materials absorb hydrogen under moderate values of pressure at low temperatures, generating reversible hydrogen compounds called hydrides. This type of hydrogen storage is often called "solid" hydrogen storage since hydrogen becomes part of the solid material through some physicochemical bonding. The use of Metal Hydrides (MH) can provide efficient solutions for hydrogen storage offering high hydrogen storage capacity per unit volume, safety and reliability, as well as high purity of the supplied H_2. Concerns related to hydrogen storage materials are the storage capacity, the thermal stability of the hydride, the kinetics of hydrogenation and dehydrogenation, the thermophysical properties, and the crystal structures. In spite of many efforts dedicated to improving the rate of the external heat transfer, the performance of the MH systems is still limited (Gkanas et al., 2016).

The hydrogen storage capacity for different storage technologies under specific temperature and pressure conditions is summarized in Fig. 4.5.

4. HYDROGEN TRANSPORTATION REQUIREMENTS

The emergence of transport systems of the hydrogen energy is the main important part of successful hydrogen economy. The transportation part will play the driven role in building a hydrogen market in different territories. It will facilitate the transport of hydrogen in territorial basis and between different territories.

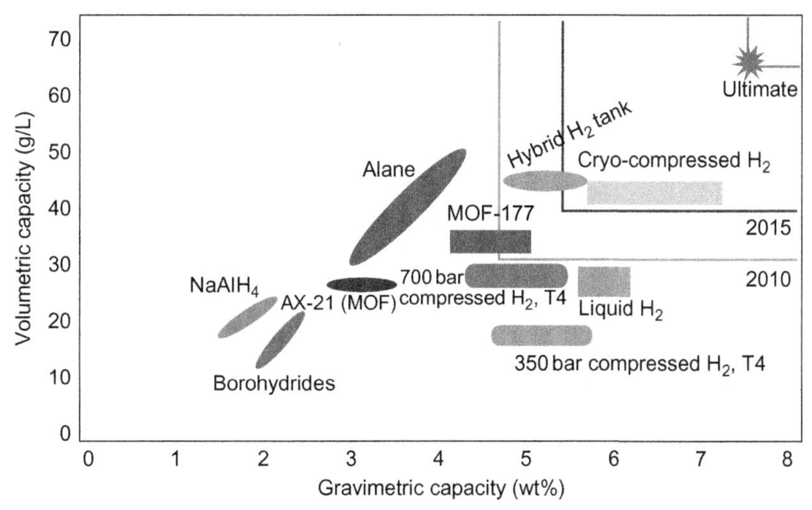

FIG. 4.5 Existing developed hydrogen storage methods with respect to DOE set requirements *(solid lines)* (Blagojević et al., 2012).

Yang and Ogden (2007) stated that the choice of the lowest-cost delivery mode (compressed gas trucks, cryogenic liquid trucks, or gas pipelines) will depend upon specific geographic and market characteristics (e.g., city population and radius, population density, size and number of refueling stations, and market penetration of fuel cell vehicles). According to Yang and Ogden (2007) and Qadrdan et al. (2008)), the main factors affecting the choice of hydrogen transport mode are the application, quantity to be transported, density of demand, and distance from the production site to the delivery points.

- Application: this factor means the type of hydrogen that is needed to be transported to the destination. For instance, if liquid hydrogen is needed for the application (liquid hydrogen refueling station), it should be delivered as liquid hydrogen (similarly in case of gaseous application where the choice will limit to gaseous transportation modes), so that, in this case, the type of the application dictates the mode of transport.
- Quantity: for large quantities, pipelines are the preferred option, especially in case of long distance, because in this case, pipeline delivery is cheaper than all other transport modes except in the case of transport over an ocean, in which liquid hydrogen transport would be the best solution. While, in case of small quantities, compressed gaseous hydrogen trucks are suitable for over short distances.
- Distance is also the deciding factor between liquid and gaseous trailers. Hydrogen transport costs are typically in the range of 1–4 ct/kWh (equivalent to 0.3–1.3 $/kg) depending on the type of deliveries and the form of hydrogen (Ball and Wietschel, 2009). Besides, for a short distance, a

pipeline might be cheaper because the capital expenditure for a short pipeline infrastructure may be close to the capital cost of tube trucks or tankers, but it doesn't includes transportation or liquefaction costs. As the distance increases, the investments for pipeline increase quickly, so that pipelines will be economically convenient for larger quantities of hydrogen to be transported.

- Density of demand: The criterion related to the density of the hydrogen demand is also key factor in the future opened hydrogen market. In fact, the concentration of the hydrogen demand will contribute to the choice of the hydrogen transportation mode and hydrogen supply infrastructures.

5. HYDROGEN TRANSPORTATION APPROACHES

5.1 Mode 1: Pipeline Transportation

Similar to the natural gas, you can think of a network of pipelines with transport functions and distribution. In fact, the gaseous hydrogen transport and distribution system might look like current natural gas pipelines with significant technological innovations: new materials for the ducts, and different working pressures and flows to overcome the reduced energy content of gaseous hydrogen (Parker, 2004).

Pipelines have been used to transport hydrogen for more than 50 years, and today, approximately 1600 miles of hydrogen pipelines are currently operating in the United States. Owned by merchant hydrogen producers, these pipelines are located where large hydrogen users, such as petroleum refineries and chemical plants, are concentrated such as the Gulf Coast region (U.S. Department of Energy. Office of Energy Efficiency and Renewable Energy, n.d.). Nearly 1600 km of hydrogen pipelines in Europe are divided between 15 larger pipeline networks with single ownerships. These main important pipeline owners in Europe are Air Liquide, Linde (BOC), and Air Products (Sapio) (European Hydrogen Infrastructure Atlas and Industrial Excess Hydrogen Analysis. Part III: Industrial Distribution Infrastructure, 2007). The largest operator in Europe, Air Liquide, has around 1300 km of hydrogen pipeline, mainly located in France, Belgium, the Netherlands, and Germany (Commission Directorate, 2006).

The longest hydrogen pipeline in the world is owned by Air Products which built 600 miles of hydrogen pipeline along America's Gulf Coast. Air Products added extra capacity for on-site facilities bringing the total system volume to over 1.4 billion SCFD (Standard Cubic Feet/Day) (over 1.5 million Nm^3/h). Customers from Texas to Louisiana can count on a stable, uninterrupted hydrogen supply (Molony, 2012). Typical operating pressures are 1–3 MPa (145–435 psig) with flows of 310–8900 kg/h (Veziroglu, 2017).

The cheapest option of transporting hydrogen is by high capacity pipeline, which can cost less than 0.1 US$/kg over 100 km (Ekins, 2010). In general, the

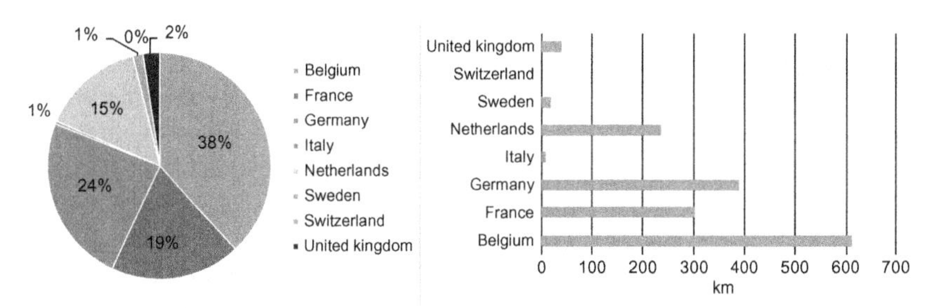

FIG. 4.6 Kilometers and percentage of hydrogen pipeline by state in Europe. *(Source: U.S. Pipeline and Hazardous Materials Safety Administration (US PHMSA State Data: Hydrogen Analysis Resource Center, n.d.).)*

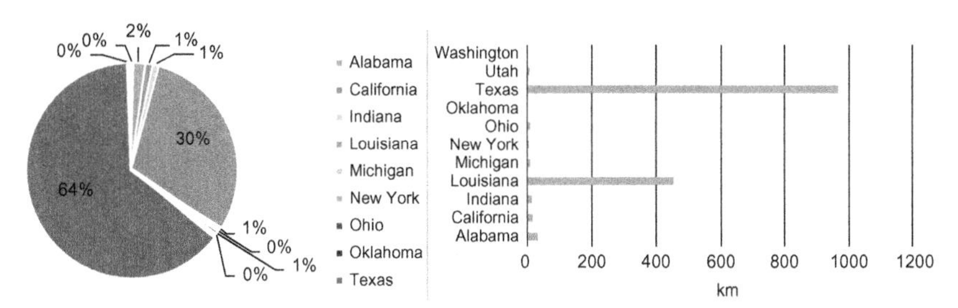

FIG. 4.7 Kilometers and percentage of hydrogen pipeline by state in USA. *(Source: U.S. Pipeline and Hazardous Materials Safety Administration (US PHMSA State Data: Hydrogen Analysis Resource Center, n.d.).)*

pipeline transport costs of hydrogen are up to 1.5 times higher in comparison to natural gas.

Figs. 4.6 and 4.7 display, respectively, the European and US pipeline hydrogen network available.

Recently, a NREL studied the possibility to deliver pure hydrogen to markets, using separation and purification technologies downstream, extracting hydrogen from the natural gas pipeline network blend close to the point of end use (Melaina and Penev, 2013). This new strategy to delivery hydrogen covers the cost of building dedicated hydrogen pipelines or other costly delivery infrastructure during the early market development phase. On the other hand, it produces additional costs, associated with blending and extraction, as well as modifications to existing pipeline integrity systems.

5.2 Mode 2: Liquid Truck

Liquid hydrogen has a high operating cost due to the electricity needed for liquefaction, but lower capital costs depending on the quantity of hydrogen and the delivery distance (Ball and Wietschel, 2009). Delivery by super-insulated, cryogenic tanker trucks is the most economical solution for medium market penetration. They could transport relatively large amounts of hydrogen and reach markets located throughout large geographic areas. Over long distances, trucking liquid hydrogen is more economical than trucking gaseous hydrogen because a liquid tanker truck can hold a much larger amount of hydrogen than a gaseous tube trailer can. In fact, a road tanker which transports high pressure hydrogen as compressed gas might typically carry 300–400kg of H_2 and be able to refuel up to about 100 cars. A tanker carrying liquefied hydrogen (LH2) can carry a much larger inventory, 2.5 to 3.5t, and so refuel about 1000 cars (Melaina and Penev, 2013). Liquid hydrogen is noncorrosive. Special materials of construction for tank trailer are not required. However, because of its extremely cold temperature, equipment must be designed and manufactured of material that is suitable for extremely low temperature operation. Vessels and piping must be selected and designed to support the pressure and temperatures involved and comply with applicable codes and regulations. ISO 13985:2006 defines the construction requirements for refillable fuel tanks used to store liquid hydrogen in land vehicles as well as the testing methods required to guarantee a reasonable level of protection from loss of life and property resulting from accident scenarios (https://www.iso.org/standard/39892.html).

Tanks are usually cylindrical in shape and placed in a horizontal position. However, some vertical cylindrical tanks and spherical tanks are in use. Standard tank sizes range from 1500 gallons to 25,000 gallons. Tanks are vacuum insulated. Pressure relief valves protect the tanks and are designed to ASME (American Society of Mechanical Engineers) specifications for the United States. The Pressure Equipment Directive (PED) 2014/68/EU (formerly97/23/EC) of the EU sets out the standards for the design and fabrication of pressure equipment ("pressure equipment" means steam boilers, pressure vessels, piping, safety valves, and other components and assemblies subject to pressure loading).

5.3 Mode 3: Tube Trailer

Trucks able to delivery gaseous hydrogen are called tube trailers. Gaseous hydrogen is compressed to pressures of 180–200 bar into long cylinders that are loaded on a trailer that the truck hauls.

Tube trailers operate currently at pressures of 250 bar by U.S. Department of Transportation (DOT) regulations. Steel tube trailers are most commonly employed and the DOT weight limitations for on-road vehicles result in a limited carrying capacity of approximately 280kg due to the heavy weight of the steel tubes. Recently, composite storage vessels have been developed that can deliver 560–720kg of hydrogen per trailer and are within DOT's height, width, and weight requirements.

The new technological solution developed by Linde group works, for example, at a higher pressure of 500 bar (7250 psi) and uses new, lighter storage materials to more than double the mass of compressed gaseous hydrogen (CGH2) that can be transported in a single truck load. A single trailer can transport over 1100 kg, or 13,000 normal cubic meters, of hydrogen gas. In addition, the trailers can now be filled and emptied in less than 60 min (Fuel Cells Bulletin, 2013).

Also Air Products launched in the UK the first of its fleet of European hydrogen high-pressure tube trailers using its SmartFuel® technology. The SmartFuel high-pressure tube trailer features specialized composite cylinders for hydrogen storage that enable cost-effective, centrally produced hydrogen to be delivered directly to fueling stations at a pressure well above 350 bar (5000 psi). This is a significant improvement on existing 200 bar (2900 psi) industrial hydrogen delivery. The increased pressure removes the need for on-site compression for 350 bar vehicle refueling and significantly reduces site compression requirements for 700 bar vehicle refueling (Fuel Cells Bulletin, 2014).

Rarely rail cars, ships, and barges are used to transport hydrogen as a CGH2 because they are not economical unless the hydrogen is in liquid form. The recent developments of tanks certified by the International Organization for Standardization (ISO tanks) that can carry large volumes of high-pressure hydrogen gas may help enable the delivery of hydrogen over rail or water.

6. CONCLUSION

This section presented a review of different hydrogen technologies that can be used in developing a future infrastructure. The review has started by presenting the main important regulations and standards related to hydrogen management. The section continued by reviewing various storage and transportation modes available for the hydrogen infrastructure.

The hydrogen-based mobility represents the main promising instrument of enabling the transport sector to meet strategic energy, environmental, and economic challenges and, in general, to manage the energy transition and move forward to a low-carbon, sustainable economy. Besides, the statistics demonstrated that people have a very positive attitude toward hydrogen-powered cars. In particular, the possibility of moving without harming the environment is appealing, but general public choice is constrained to the condition that the hydrogen has to be produced using renewable energy resources (Zimmer and Welke, 2012).

An issue that confronts the use of high-pressure and cryogenic storage centers is public perception and acceptability associated with the use of pressurized gas and liquid hydrogen containment. In fact, large studies have been devoted to investigate the issues related to risks of using gaseous and liquid hydrogen. Hydrogen storage is regarded as one of the most critical issues, which must be solved before a technically and economically viable hydrogen infrastructure implementation. In fact, without effective storage systems, a hydrogen economy will be difficult to achieve.

In 2011, authors in (Ogden and Nicholas, 2011) analyzed a cluster strategy for introducing FCVs in Southern California, U.S. The cluster strategy proposed to construct H_2 infrastructure in a few focused geographic areas. It states that once the hydrogen demand is sufficient to support fully utilized 1000 kg/day stations, hydrogen could be produced at $5–7/kg.

In 2016, the London's first hydrogen filling station refueled a car in 3 min filling a tank with 5 kg of pressurized hydrogen at a cost of £10/kg, giving a range of between 300 and 500 miles, depending on model.

The NREL (National Laboratory of the U.S. Department of Energy, Office of Energy Efficiency and Renewable Energy) developed the Hydrogen Financial Analysis Scenario Tool, H2FAST (http://www.nrel.gov/hydrogen/h2fast/), which is able to provide a quick and convenient in-depth financial analysis for hydrogen fueling stations.

The user can access to the system and insert data for three different types of input. The station inputs define the user's model hydrogen station: installation time, station type (delivered gaseous H_2, delivered liquid H_2, on-site H_2 production by electrolysis or by SMR), total hydrogen capacity, and installation costs. The scenario inputs are related to incentives; incidental revenue; and hydrogen, electricity, and natural gas costs. Financial inputs concern debt interest rate for the investments. The H2FAST tool provides users with detailed annual finance projections in the form of income statements, cash flow statements, and other financial performance parameters related to the realization of hydrogen filling stations.

REFERENCES

Air products hydrogen high-pressure tube trailers for Europe. Fuel Cells Bull. 2014 (2), 7–8.

Ball, M., Wietschel, M., 2009. The future of hydrogen—opportunities and challenges. Int. J. Hydrog. Energy 34 (2), 615–627.

Barbier, F., 2010. Hydrogen distribution infrastructure for an energy system: present status and perspectives of technologies.World Hydrogen Energy Conference 2010 May 16–21, Messe Essen, Germany.

Barthelemy, H., Weber, M., Barbier, F., 2017. Hydrogen storage: recent improvements and industrial perspectives. Int. J. Hydrog. Energy 42 (11), 7254–7262.

Blagojević, V.A., Minić, D.G., Grbović Novaković, J., Minic, D.M., 2012. In: Minic, D. (Ed.), Hydrogen Economy: Modern Concepts, Challenges and Perspectives, Hydrogen Energy—Challenges and Perspectives. InTech. https://doi.org/10.5772/46098. Available from: https://www.intechopen.com/books/hydrogen-energy-challenges-and-perspectives/hydrogen-economy-modern-concepts-challenges-and-perspectives.

Commission Directorate, 2006. Commission Directorate-General Environment Assessing the Case for EU Legislation on the Safety of Pipelines and the Possible Impacts of Such an Initiative. Final Report ENV.G.1/FRA/2006.

Das, L.M., Gulati, R., Gupta, P.K., 2000. A comparative evaluation of the performance characteristics of a spark ignition engine using hydrogen and compressed natural gas as alternative fuels. Int. J. Hydrog. Energy 25 (8), 783–793.

Durbin, D.J., Allan, N.L., Malardier-Jugroot, C., 2016. Molecular hydrogen storage in fullerenes—a dispersion-corrected density functional theory study. Int. J. Hydrog. Energy 41 (30), 13116–13130.

Ekins, P., 2010. Hydrogen Energy: Economic and Social Challenges. Earthscan Publication, London.

European Commission, 2006. Introducing Hydrogen as an Energy Carrier Safety, Regulatory and Public Acceptance Issues. Results from EU Research Framework Programmes. Directorate-General for Research Sustainable Energy Systems.

Steinberger-Wilckens, R., Trümper, S.C. (Eds.), 2007. European Hydrogen Infrastructure Atlas and Industrial Excess Hydrogen Analysis. Part III: Industrial Distribution Infrastructure. PLANET GbR, Oldenburg, Germany. Document Tracking No.: R2H2001PU, https://www.ika.rwth-aachen.de/r2h/index.php/European_Hydrogen_Infrastructure_and_Production.html. Accessed May 2017.

Gkanas, E.I., Grant, D.M., Khzouz, M., Stuart, A.D., Manickam, K., Walker, G.S., 2016. Efficient hydrogen storage in up-scale metal hydride tanks as possible metal hydride compression agents equipped with aluminium extended surfaces. Int. J. Hydrog. Energy 41 (25), 10795–10810.

Global Technical Regulation. Hydrogen fueled vehicle: proposal to develop a global technical regulation concerning hydrogen fuel cell vehicle (ECE/TRANS/WP.29/AC.3/17).

He, C., Yu, R., Sun, H., Chen, Z., 2016. Lightweight multilayer composite structure for hydrogen storage tank. Int. J. Hydrog. Energy 41 (35), 15812–15816.

Hosseini, M., Dincer, I., Naterer, G.F., Rosen, M.A., 2012. Thermodynamic analysis of filling compressed gaseous hydrogen storage tanks. Int. J. Hydrog. Energy 37 (6), 5063–5071.

HySafeVersion 1.0, 2006. Chapter V: Hydrogen Safety Barriers and Safety Measures. http://www.hysafe.org/download/1200/BRHS_Chap5_V1p2.pdf.

Linde Group, The cleanest energy carrier ever. Hydrogen Solutions from Linde Gas. http://www.the-linde-group.com/internet.global.thelindegroup.global/en/images/HydrogenBrochure_EN14_10196.pdf?v.

Linde raises pressure for hydrogen transport efficiency, fills in US. Fuel Cells Bull. 2013 (10), 7.

Melaina, O.A., Penev, M., 2013. Blending Hydrogen into Natural Gas Pipeline Networks: A Review of Key Issues M. W. NREL. National Laboratory of the U.S. Department of Energy, Office of Energy. Technical Report NREL/TP-5600-51995.

Molony, J., 2012. Air products dedicates world's largest hydrogen pipeline system. In: Pasadena Citizen Newspaper. October 28.

Ogden, J., Nicholas, M., 2011. Analysis of a "cluster" strategy for introducing hydrogen vehicles in Southern California. Energ. Policy 39, 1923–1938.

Parker, N., 2004. Using natural gas transmission pipeline costs to estimate hydrogen pipeline costs. Research Report, Institute of Transportation Studies, University of California, Davis.

Prabhukhot, P.R., Wagh, M.M., Gangal, A.C., 2016. A review on solid state hydrogen storage material. Adv. Energy Power 4 (2), 11–22.

Qadrdan, M., Saboohi, Y., Shayegan, J., 2008. A model for investigation of optimal hydrogen pathway, and evaluation of environmental impacts of hydrogen supply system. Int. J. Hydrog. Energy 33 (24), 7314–7325.

Roads2HyCom. Report on assessment of hydrogen safety & security (S&S) and hydrogen regulations. Document Number: R2H1009PU.2, 2008. https://www.ika.rwth-aachen.de/r2h/index.php/Roads2HyCom_Reports_in_Detail.html (accessed May 2017).

U.S. Department of Energy, Fuel Cell Technologies Office. http://hydrogen.pnl.gov/hydrogen-data/hydrogen-properties. Accessed April 2017.

U.S. Department of Energy. Office of Energy Efficiency and Renewable Energy, Hydrogen Pipelines. Available at: https://energy.gov/eere/fuelcells/hydrogen-pipelines. Accessed May 2017.

U.S. Drive, 2013. Hydrogen Delivery Technical Team Roadmap. https://energy.gov/sites/prod/files/2014/02/f8/hdtt_roadmap_june2013.pdf.

US PHMSA State Data: Hydrogen Analysis Resource Center. http://hydrogen.pnl.gov/hydrogen-data/hydrogen-delivery.

Veziroglu, A., 2017. Hydrogen Powered Transportation. Xlibris Corporation.

Yang, C., Ogden, J., 2007. Determining the lowest-cost hydrogen next term delivery mode. Int. J. Hydrog. Energy 32, 268–286.

Zheng, J., Liu, X., Xu, P., Liu, P., Zhao, Y., Yang, J., 2012. Development of high pressure gaseous hydrogen storage technologies. Int. J. Hydrog. Energy 37 (1), 1048–1057.

Zimmer, R., Welke, J., 2012. Let's go green with hydrogen! The general public's perspective. Int. J. Hydrog. Energy 37 (22), 17502–17508.

Zuttel, A., 2004. Hydrogen storage methods. Naturwissenschaften 91, 157–172.

FURTHER READING

Hacker, V., Nestl, S., Friedrich, T., 2016. In: Progress in hydrogen production and storage technologies.2016 International Conference on the Domestic Use of Energy (DUE), Cape Town, pp. 1–6.

Lowesmith, B.J., Hankinson, G., Chynoweth, S., 2014. Safety issues of the liquefaction, storage and transportation of liquid hydrogen: an analysis of incidents and HAZIDS. Int. J. Hydrog. Energy 39 (35), 20516–20521.

Xu, P., Zheng, J., Liu, P., Chen, R., Kai, F., Li, L., 2009. Risk identification and control of stationary high-pressure hydrogen storage vessels. J. Loss Prev. Process Ind. 22 (6), 950–953.

Chapter 5

Deployment of a Hydrogen Supply Chain

1. CHALLENGES FOR THE DESIGN OF THE FUTURE HYDROGEN SUPPLY CHAIN (HSC)

Hydrogen is one of the most promising alternatives for future transportation fuel. According to IEA energy technology essentials (2007), hydrogen can gain significant market share over the coming decades if the cost of hydrogen production, distribution, and end use can fall significantly and if effective policies are put in place to increase energy efficiency, mitigate CO_2 emissions, and improve energy security. In addition, hydrogen is a key enabler for more widespread implementation of sustainable technologies and more efficient products in the marketplace.

However, infrastructure issues pose more challenges and uncertainties for hydrogen. In particular, the choice of optimal configuration of the HSC is complicated by the consideration that many alternatives are available for each node of the chain (production, storage, transportation, end users). The development of global hydrogen pathway faces usually the need to solve the compromise between many vital criteria. For instance, some solutions that could be beneficial from an economic viewpoint are usually unsustainable from the environmental viewpoint. According to a time basis, some infrastructure options could have increasing role in a long term, but unavailable in the initial phases due to some technical and economic barriers.

Several valuable contributions in the design of HSC have been presented in the previous sections. The literature review reveals that research over the last decades has focused on many aspects of the HSC.

The studies published in the literature range from a complete design of the global HSC to a focus on specific nodes. Strategies for designing the hydrogen economy are established based on careful analysis that takes into account critical issues such as cost, environment, and safety. It is recognized according to the literature review that models, methods, and approaches for the planning of future HSC are mainly focusing on the mathematical optimization. These research studies have tried to find the optimal configuration of HSC optimizing a unique criterion or multicriteria. One common point between papers based on mathematical optimization lies in the cost minimization of HSC. Some authors

Hydrogen Infrastructure for Energy Applications. https://doi.org/10.1016/B978-0-12-812036-1.00005-6
© 2018 Elsevier Inc. All rights reserved.

have focused on the minimization of various costs related to the node of the HSC, while others have focused on the minimization of the environmental impacts of the HSC. Fewer studies have addressed the optimization of the HSC from the risks viewpoint. These criteria may be of highly interest taking into account the particularity of hydrogen. Also, from the production viewpoint, there is a need to investigate HSC that operates on clean feedstock, such those based on renewable energy resources. Future research papers are also needed to cover the technical aspect related to the operation of the HSC. Focus must be dedicated to the evaluation of the technical feasibility and the performance of renewable HSC. In addition, to encourage the widespread use of hydrogen economy, comparison of HSC with the conventional network of petrol products should be done, which could help in decreasing some uncertainties along the future HSC and succeed to the commercialization of the hydrogen as a fuel. Even though greatest numbers of research studies have been accomplished worldwide, it must be borne in mind that a successful transition to a hydrogen economy cannot be guaranteed that easy. The development of new models, equipments, and other technical standards represents serious challenges for the commercialization of hydrogen. Cost reduction of hydrogen production and cost-competitive transportation are considered one of the major defies to the success of hydrogen infrastructures. More attention should be given to hydrogen production and delivery challenges like lowering cost of hydrogen production from RES, offering secure production that can be considered as sustainable. The adoption of new strategies and initiatives, policies, government hydrogen research subventions, and specific hydrogen programs will promote and advance the use and acceptance of hydrogen as a fuel. In fact, more support is needed from the national governments. In turn, government preferences must be included in the modeling and the design phase of HSC. New perceptions of hydrogen demand market are needed to better understand the future hydrogen infrastructure. This can be done through including new estimation methods of hydrogen demand that will mainly come from real market.

2. HYDROGEN RENEWABLE ENERGY RESOURCES: ANALYSIS AND MODELING

2.1 Introduction

Renewable energy is an environmental friendly option which may be economically competitive with conventional power generation, where good resources are available. It represents an opportunity to enhance sustainable development in both countries which traditionally lacks fossil fuels and those who are constrained by some environmental policies and regulations. Renewable energy can play increasing role in countries that have an important environmental wealth and great renewable energy potential. Quality and accessibility of resource data will enable private investors and public policy makers to access the technical, economic, and environmental potential for large-scale investments in green technologies. For more accurate resource assessment, detailed site-specific micro-siting analysis should be done. Renewable power generation has shown a remarkably rapid growth in the past twenty years, and now it is a mature, reliable, and efficient technology. In addition, in regions with proper resource characteristics, renewable energy may already be competitive with traditional fuels.

Any choice of renewable energy sources exploitation site must be based on the preliminary investigation of the average wind velocity, solar irradiation, and potential, so that the accuracy of the resources' data analysis is a crucial factor to be undertaken.

In the literature, many studies have been focused on providing a forecasting tool to predict and assess renewable energy sources and power production with good accuracy. From the wind speed assessment viewpoint, several authors (Rehman, 2004; Shata and Hanitsch, 2008; Himri et al., 2009; Celik, 2006; Ucar and Balo, 2009; Conte et al., 1998) have used many approaches to evaluate wind speed and wind energy production. This chapter aims to investigate the wind speed and solar irradiation characteristics of the Liguria region (Italy). The study is performed following two steps. In the first step, the study is based on the analysis and modeling of the statistical data; in this step, attention is given also to a deep analysis of wind sites that have shown promising potential for future exploitation. In the second step, a GIS tool, more specifically, an ArcGIS Spatial Analyst is used to produce maps of wind and solar energy.

2.2 Wind Speed and Solar Irradiation Data

The estimation of a potential is based on the knowledge of wind and solar regimes on the considered territory. The assessment of the Liguria Region wind and solar potential has been carried out using data gathered by the Agenzia Regionale della Protezione Ambientale Liguria (ARPAL). In total, data from 25 stations for wind and 16 stations for solar distributed over the four provinces of Liguria (Figs. 5.1 and 5.2) (i.e., La Spezia, Genoa, Savona, and Imperia) have been analyzed. It should be pointed out that different periods of monitoring are available for different sites; these data were the only ones that are available and which have been gathered on-site by the regional agency for environmental protection in the Liguria region.

The wind data used in this current study were observed in the following locations:

- Casoni (2002–June/2008),
- Borgonuovo, Castellari, Cavi di Lavagna, Cenesi, Imperia, Levanto-s Gottardo, Monte Rocchetta and Polanesi (2003–June/2008)
- Genova villa Cambiaso (2002–2006)

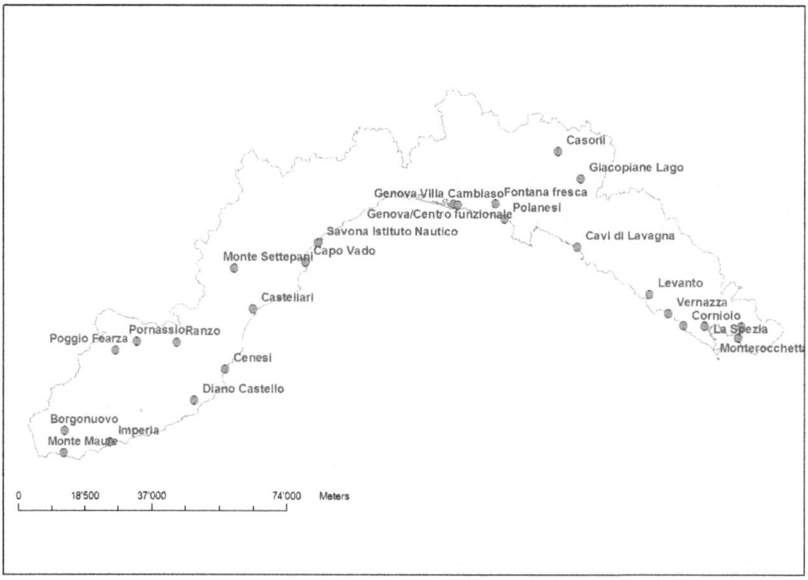

FIG. 5.1 The geographical locations of wind meteorological stations.

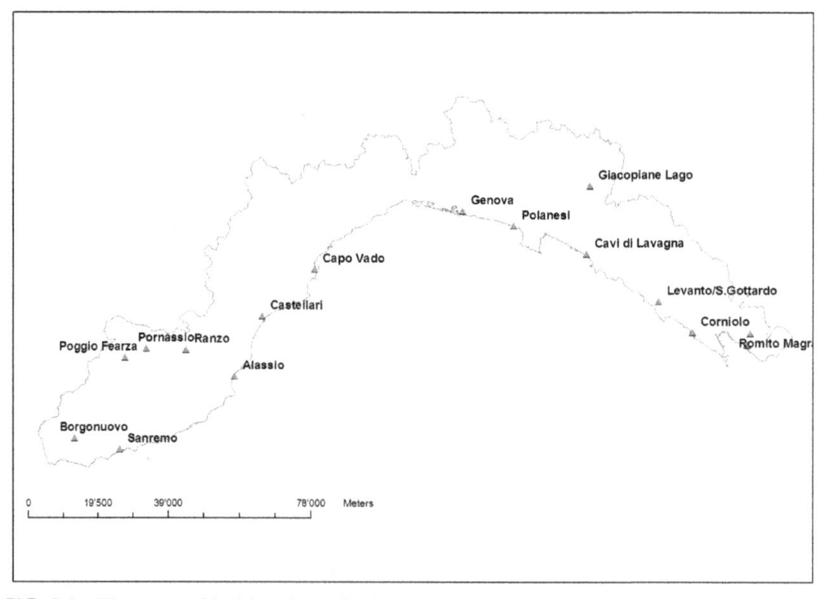

FIG. 5.2 The geographical locations of solar meteorological stations.

- Diano Castello (2003–2007)
- Monte Settepani, Fontana Fresca, Monte Maure and Ranzo (2004–June/2008),
- Vernazza (2004–2007)
- Giacopiane Lago and Pornassio (2005–June/2008)
- Romito Magra (2004–2006)
- Capo Vado (2006–June/2008)
- Genova, ARPAL functional center (2007–June/2008)
- Savona (April/2007–June/2008), Poggio Fearza (2007)
- Corniolo and La Spezia (2008–June/2008).

The solar data used in this current study were observed in the following locations:

- Castellari (2008 Jan/2009 Dec)
- Genova (2005-Jan/009-Dec)
- Polanesi (2006-Jan/2009-Dec)
- Monterocchetta
- Corniolo (2009-Jan/2009-Dec)
- Buorgonuovo (2005-Jan/2009-Dec)
- Capo Vado (2008-Jan/2009-Dec)
- Cavi di Lavagna (2005-Jan/2009-Dec)
- Levanto\ S.Gottardo
- Allassio (2007-Jan/2009-Dec)
- Sanremo (2008-Jan/2009-Dec)
- Romito Magra (2006-Jan/2009-Dec)
- Ranzo (2006-Jan/2009-Dec)
- Pornassio (2006-Jan/2009-Dec)
- Giacopiane Lago
- Poggio Fearza (2009-Jan/2009-Dec)

The data have been used to evaluate the annual frequency of wind speed, the monthly and annual variations as regards average speed, the vertical profile of the wind speed, and the assessment of wind and solar power potential.

2.3 Wind Energy Analysis Model

The computation of the wind speed probability distribution function (PDF) constitutes the first fundamental step to assess the wind energy potential, since it can effectively determine the performance of wind energy systems for a given location and time (Kantar and Usta, 2008; Carta et al., 2009). Several PDFs have been proposed in the literature to represent the frequencies of the wind speed. The Weibull with its two characteristic parameters is the most commonly used and different estimation methods can be used for its identification.

The general form of the Weibull PDF is:

$$f(v) = \left(\frac{k}{c}\right)\left(\frac{v}{c}\right)^{k-1} \exp\left(-\left(\frac{v}{c}\right)^{k}\right) \tag{5.1}$$

where $f(v)$ is the probability of observing wind speed v, k is the dimensionless Weibull shape parameter, and c is the Weibull scale parameter.

According to the measurements of wind speed at a specific site, it is necessary to use estimation methods to derive parameters k and c on the basis of the known data v. Common estimation methods can be applied: the standard deviation method (SD), the maximum likelihood method (MLM), or the least squares method (LSM). In this study, the SD method has been used, so that the two parameters of Weibull PDF, k and c can be related to the mean speed V_m and standard deviation σ by (Ouammi et al., 2010):

$$V_m = \int_0^\infty v f(v) dv = \int_0^\infty v \frac{k}{c}\left(\frac{v}{c}\right)^{k-1} e^{-\left(\frac{v}{c}\right)^K} dv = c\Gamma\left(1 + \frac{1}{k}\right) \tag{5.2}$$

where,

$$\Gamma(y) = \int_0^\infty e^{-x} x^{y-1} dx \tag{5.3}$$

$$k = \left(\frac{\sigma}{V_m}\right)^{-1.086} \tag{5.4}$$

$$c = \frac{V_m}{\Gamma\left(1 + \frac{1}{k}\right)} \tag{5.5}$$

2.3.1 Extrapolation of Data at Hub Height

The wind speed data are collected at a height H_{data} (m) that is different from the height of the hub. So, it is necessary to represent the relation among wind speed v_{hub} (m/s) at hub height H_{hub} (m), the wind speed v_{data} (m/s) at H_{data}, and the surface roughness length z_0 (m). In this work, a relation proposed in Ouammi et al. (2010) is used, namely,

$$v_{hub} = v_{data} \frac{\ln(H_{hub}/z_0)}{\ln(H_{data}/z_0)} \tag{5.6}$$

2.3.2 Wind Power Density

The power of the wind that flows at speed v through the blade sweep area A (m^2) increases as the cubic of its velocity and is given by (Ouammi et al., 2010):

$$P(v) = \frac{1}{2}\rho A v^3 \tag{5.7}$$

where ρ (kg/m^3) is the density of air.

The wind power density of a site based on Weibull's probability density function can be expressed as follows (Ouammi et al., 2010):

$$P = \int_0^\infty P(v) f(v) dv = \frac{1}{2} A\rho c^3 \Gamma\left(\frac{k+3}{k}\right) \tag{5.8}$$

2.3.3 Classes of Wind Power Density

The Battelle-Pacific Northwest Laboratory (PNL) developed a wind power density classification scheme to classify the wind resources. The Battelle-PNL classification is a numerical one which includes rankings from Wind Power Class 1 (lowest) to Wind Power Class 7 (highest). Each class represents a range of wind power density (W/m^2) or a range of equivalent mean wind speeds at specified heights above ground level (Ilinca et al., 2003). Class 4 or greater is considered to be suitable for most wind turbine applications. Class 3 areas are suitable for wind energy development using taller wind turbine towers. Class 2 areas are considered marginal for wind power development and Class 1 areas are unsuitable (Ilinca et al., 2003). More description of the Battelle-PNL classification can be found in Ilinca et al. (2003).

As determined by Ouammi et al. (2010), taking into account a statistical analysis on the whole year, the wind power densities of Capo Vado, Casoni, Fontana Fresca, and Monte Settepani are equal, respectively, to 487.7, 332.5, 206.5, and 203 W/m^2. Consequently, Capo Vado appears to have potential wind resources as Class 7, Casoni is classified as Class 6, and both Fontana Fresca and Monte Settepani are classified as wind power Class 4.

2.4 Modeling the Wind Power Plant and Its Performance

The estimation of the exploitable energy requires the definition of the performances of the wind power plant (WPP) system. The WPP simplified model which is taken into account in this work is related to a horizontal axis wind turbine equipped with a gearbox. So, the WPP is supposed to consist of three main components: the rotor R, the gearbox GB, and the generator G.

The site will be characterized by a wind speed v (m/s) with a statistical distribution function $f(v)$, equipped with a WPP whose efficiency is C_{WPP} and with a rotor sweeping a surface A (m^2), working in a range of wind speed $v \in [v_i, v_f]$, where v_i (m/s) is the cut-in wind speed, and v_f (m/s) is the cut-off wind speed. Under the above-mentioned hypothesis, the electric energy E_{wt} (kWh) which can be produced per time period T is given by (Ouammi et al., 2012):

$$E_{wt} = \frac{T}{1000} \frac{\rho}{2} A \int_{v_i}^{v_f} v^3 f(v)_{hub} C_{WPP}(v) dv \tag{5.9}$$

where T is the number of hours, ρ (kg/m^3) is the air density, $f(v)_{hub}$ is the Weibull PDF at the height of the hub, and A (m^2) is the area swept by the WPP blades.

The WPP performance coefficient of the plant system, $c_{WPP}(v)$, is made by three related components, which are also dependent on the wind speed v (Ouammi et al., 2012):

$$c_{WPP}(v) = c_P(v) \eta_{GB}(v) \eta_G(v) \tag{5.10}$$

2.5 Case Study

The obtained results from the available meteorological data used in this study show that in some sites the wind energy potential is low, while in other sites wind potential is considerably high. With the aim of covering all four provinces of Liguria, we used data from all the available stations, even if most of them are not located in the windiest parts of the territory. The average wind speed (on the whole measurement period) has been evaluated for each site and presented in Table 5.1. The results show that Capo Vado (6.52 m/s), Casoni (5.74 m/s), Monte Settepani (5.45 m/s), and Fontana Fresca (5 m/s), have the highest wind speed at 10 m. Unlike other sites, among these four sites (Table 5.2) Capo Vado is considered the most promising site for the Liguria Region with an available energy of 4271.7 kWh/m^2 yr.

Table 5.3 presents the annual variation of average wind speed of each of these sites. For Capo Vado, a minimum average speed of 6.12 m/s in 2006 and a maximum of 6.87 m/s in 2008 have been monitored, which show that the average wind speed of the site has not undergone considerable changes and has kept the same order of magnitude. The same conclusion can be drawn for Casoni with a minimum of 5.73 m/s in 2008 and a maximum of 6.12 m/s in 2006, and for Monte Settepani with a minimum of 5.16 in 2008 and a maximum of 5.79 m/s in 2007. Similar results are obtained for Fontana Fresca with a minimum of 4.94 m/s in 2004 and a maximum of 5.04 m/s in 2008.

ArcGIS Spatial Analyst has been implemented to display the distribution of the wind and solar in the region. The ArcGis tool includes specific interpolation methods to estimate spatially continuous phenomena, here it is related to solar and wind. The aim of the interpolation methods is to create a surface grid in ArcGIS in order to assess the values of cells at locations that lack sampled points. Among interpolation tools, there are spline, kriging methods, and

TABLE 5.1 The average wind speed on the whole measurements period for each site

Sites	Wind speed (m/s) 10 (m)	Wind speed (m/s) 40 (m)	*k* (SD)	c (m/s) (SD)
Borgonuovo	1.15	1.86	2.3	1.3
Capo Vado	6.52	8.56	1.43	7.18
Castellari	1.28	2.06	1.10	1.32
Cavi di Lavagna	1.25	2.03	0.98	1.24
Cenesi	1.55	2.42	1.43	1.70
Corniolo	2.76	4.05	1.09	2.85
Fontana Fresca	5.00	6.76	1.48	5.51
Genova Centro	3.64	5.14	1.29	3.94
Genova Villa Cambiaso	1.90	2.89	1.63	2.12
Giacopiane Lago	3.73	5.28	1.06	3.82
La Spezia	3.15	4.51	1.43	3.47
Levanto-S Gottardo	1.48	2.32	1.46	1.63
Monte Maure	3.86	5.38	1.69	4.32
Monte Rocchetta	3.80	5.33	1.48	4.21
Monte Settepani	5.45	7.29	1.84	6.13
Poggio Fearza	4.72	6.44	1.51	5.23
Polanesi	1.21	1.95	2.31	1.37
Pornassio	1.24	1.98	1.92	1.39
Ranzo	2.35	3.48	2.23	2.65
Romito Magra	1.01	1.66	1.38	1.10
Savona Istituto Nautico	3.47	4.90	2.44	3.91
Vernazza	1.65	2.55	1.84	1.85
Casoni	5.74	7.30	1.43	6.34
Diano Castello	0.69	1.20	0.99	0.69
Imperia	3.35	4.77	1.54	3.72

TABLE 5.2 Available wind energy for each site

Sites	k	c (m/s)	Power density (W/m^2)	Available energy (kWh/m^{2}yr)	Altitude (m)	Latitude N	Longitude E
Borgonuovo	2.3	1.3	1.54	13.6	100	43.8463	7.6208
Capo Vado	1.43	7.18	487.7	4271.7	170	44.2583	8.4425
Castellari	1.10	1.32	6.04	52.9	100	44.1456	8.2625
Cavi di Lavagna	0.98	1.24	7.5	65.7	100	44.2961	9.3739
Cenesi	1.43	1.70	6.47	56.7	110	44.0750	8.1347
Corniolo	1.09	2.85	62	543.5	258	44.1063	9.7348
Fontana Fresca	1.48	5.51	206.5	1809.3	743	44.4022	9.0936
Genova Centro	1.29	3.94	100	874.3	20	44.4017	8.9472
Genova Villa	1.63	2.12	9.90	86.7	40	44.3986	8.9633
Giacopiane Lago	1.06	3.82	161.75	1417	1016	44.4608	9.3875
La Spezia	1.43	3.47	54.47	477.2	5	44.1045	9.8075
Levanto-S Gottardo	1.46	1.63	5.5	48	100	44.1811	9.6211
Monte Maure	1.69	4.32	80	694.5	210	43.7922	7.6192
Monte Rocchetta	1.48	4.21	92	805.2	412	44.0755	9.9197
Monte Settepani	1.84	6.13	203	1776.5	1375	44.2430	8.1966
Poggio Fearza	1.51	5.23	170	1483.5	1833	44.0420	7.7935

Polanesi	2.31	1.37	1.8	15.8	50	44.3658	9.1247
Pornassio	1.92	1.39	2.3	19.8	500	44.0639	7.8664
Ranzo	2.23	2.65	13.5	118	310	44.0632	8.0049
Romito Magra	1.38	1.10	2	16.9	100	44.1033	9.9303
Savona Istituto Nautico	2.44	3.91	40.23	352.4	38	44.3056	8.4855
Vernazza	1.84	1.85	5.61	49.2	160	44.1361	9.6833
Casoni	1.43	6.34	332.5	2908.8	800	44.5272	9.3086
Diano Castello	0.99	0.69	1.20	10.5	135	43.9232	8.0669
Imperia	1.54	3.72	59	516.7	10	43.8882	8.0416

TABLE 5.3 Annual average wind speed (a), annual power density (b), and annual available energy for the four sites with highest wind speed (c)

	Capo Vado	Casoni	Fontana Fresca	Monte Settepani
(a) Annual average wind speed (m/s)				
2004	–	5.82	4.94	5.30
2005	–	5.73	4.94	5.60
2006	6.12	6.12	5.03	5.40
2007	6.56	5.84	4.99	5.79
2008	6.87	5.97	5.04	5.16
(b) Power density (W/m^2)				
2004	–	355.98	183.80	194.02
2005	–	366.41	200.16	197.92
2006	367.58	367.59	211.40	192.09
2007	483.58	329.03	202.04	222.72
2008	620.71	374.64	237.08	207.54
(c) Available wind energy (MWh/m^2 yr)				
2004	–	3.12	1.61	1.70
2005	–	3.21	1.75	1.73
2006	3.22	3.22	1.85	1.68
2007	4.24	2.88	1.77	1.95
2008	5.44	3.28	2.08	1.82

inverse distance weighted (IDW). The most appropriate method depends on the distribution of the sampled points as the set of points is dense enough to capture the extent of surface variation. For this reason, this method has been excluded for the case of Liguria region. We should mention that different methods have been tested to evaluate the phenomena and create surface grid, the kriging method is the one that gave the most appropriate results. The data from all the sites have been given as input to a Kriging algorithm in order to produce available wind energy at 10 m.

From Fig. 5.3, it is quite evident that different sites of Liguria region have very different wind potential characteristics. It seems that some internal territories on the mountains as well as some part of the coast in the western side are more promising than others for the exploitation of the wind resource for energy production. In addition, this potential seems quite stable in the years.

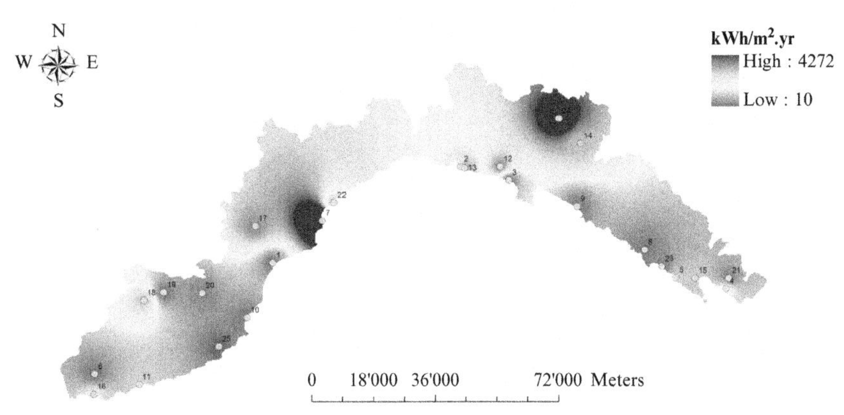

FIG. 5.3 Available wind energy in Liguria region.

However, as usual in these cases, also due to the complex orography of Liguria region, a monitoring campaign on the field should be performed on the site. From the data obtained on the 25 stations, only 4 of them seem to be eligible for energy production.

The WPP with the following geometric and technical characteristics has been considered to assess the energy output: rated power 1500kW, cut-in wind speed of 4m/s, rated wind speed of 14m/s, cut-off wind speed of 20m/s, survival wind speed of 52.5m/s, 3 blades, diameter 82m, hub height 76m, and swept area of 5281 m^2. Furthermore, the power coefficient, gearbox efficiency, and generator efficiency have been supposed equal, respectively, to 0.45, 0.96, and 0.96.

The wind speed data of the four locations (Capo Vado, Casoni, Fontana Fresca, and Monte Settepani) have been analyzed taking into account the monthly and seasonal variations. The monthly variation of Weibull parameters (k and c) and the mean monthly wind speed at 10 and 76 m above the ground level are listed in Tables 5.4–5.7.

For Capo Vado site (Table 5.5), it can be observed that the maximum value of the monthly mean wind speed at 10m is 9.98m/s in December and a minimum value of 2.80m/s occurs in April, while at the hub height the monthly wind speed varies between 3.73 and 13.46m/s. Furthermore, Weibull shape parameter k varies between 0.95 and 1.94, while scale parameter c between 2.73 and 11.25m/s.

Results of Fontana Fresca location (Table 5.6) reveal that at 10m, a value of 6.78m/s is observed as a maximum monthly wind speed in December and a minimum value of 3.69m/s in June, furthermore at the hub height, the monthly wind speed is ranging between 4.98 and 9.15m/s. The shape parameter k varies between 1.36 and 1.92 while the scale parameter c between 4.15 and 7.46m/s.

As concern Monte Settepani (Table 5.7), it appears that December is still the month that gives the maximum value of the wind speed, and which attains

TABLE 5.4 Monthly variations of the mean wind speed and Weibull parameters in Capo Vado site

Month	c (m/s)	*k*	Mean wind speed (m/s) 10 (m)	Mean wind speed (m/s) 76 (m)	Power density (W/m^2)
January	7.50	1.36	6.87	9.27	620.14
February	7.15	1.63	6.40	8.63	381.04
March	7.05	1.46	6.38	8.61	441.65
April	6.21	1.30	5.73	7.73	384.75
May	5.42	2.03	4.80	6.48	125.21
June	5.53	1.89	4.91	6.62	143.73
July	5.13	1.86	4.56	6.15	117.11
August	4.97	1.83	4.41	5.95	108.32
September	5.61	1.54	5.05	6.81	202.53
October	6.51	1.48	5.88	7.94	339.62
November	7.04	1.44	6.39	8.62	452.95
December	8.62	1.39	7.86	10.61	894.86

TABLE 5.5 Monthly variations of the mean wind speed and Weibull parameters in Casoni site

			Capo Vado		
Month	c *(m/s)*	k	*Mean wind speed (m/s) 10 (m)*	*Mean wind speed (m/s) 76 (m)*	*Power density (W/m^2)*
January	5.91	1.45	5.36	7.24	265.52
February	7.60	1.64	6.80	9.17	453.93
March	7.80	1.53	7.02	9.47	550.20
April	2.73	0.95	2.80	3.78	90.71
May	5.06	1.36	4.63	6.24	187.71
June	5.29	1.58	4.75	6.41	162.55
July	4.95	1.53	4.46	6.01	140.21
August	5.78	1.82	5.14	6.93	172.04
September	6.42	1.49	5.80	7.83	321.06
October	9.22	1.89	8.19	11.05	666.03
November	9.06	1.62	8.11	10.95	783.07
December	11.25	1.94	9.98	13.46	1177.97

TABLE 5.6 Monthly variations of the mean wind speed and Weibull parameters in Fontana Fresca site

Fontana Fresca					
Month	c *(m/s)*	k	*Mean wind speed (m/s) 10 (m)*	*Mean wind speed (m/s) 76 (m)*	*Power density (W/m^2)*
January	6.65	1.73	5.93	7.99	279.36
February	5.92	1.47	5.36	7.22	257.13
March	6.61	1.68	5.90	7.96	286.11
April	4.52	1.53	4.07	5.49	107.03
May	5.10	1.82	4.53	6.12	117.81
June	5.15	1.92	4.57	6.17	114.48
July	4.15	1.76	3.69	4.98	66.13
August	4.86	1.68	4.34	5.86	113.80
September	4.52	1.41	4.11	5.55	124.44
October	5.64	1.36	5.17	6.97	261.95
November	6.13	1.50	5.53	7.46	275.45
December	7.46	1.43	6.78	9.15	547.46

TABLE 5.7 Monthly variations of the mean wind speed and Weibull parameters in Monte Settepani site

Monte Settepani					
Month	c *(m/s)*	k	*Mean wind speed (m/s) 10 (m)*	*Mean wind speed (m/s) 76 (m)*	*Power density (W/m^2)*
January	7.04	2.26	6.24	8.41	248.60
February	7.58	2.16	6.72	9.06	322.22
March	6.30	1.90	5.59	7.55	212.04
April	6.28	2.45	5.57	7.51	166.26
May	5.81	2.34	5.15	6.95	136.11
June	5.61	2.43	4.97	6.71	118.76
July	5.30	2.25	4.69	6.33	106.21
August	5.70	2.11	5.05	6.81	140.18
September	5.88	2.07	5.21	7.03	157.08
October	7.12	1.95	6.32	8.52	296.46
November	7.07	1.95	6.27	8.46	289.53
December	7.83	2.32	6.94	9.36	334.71

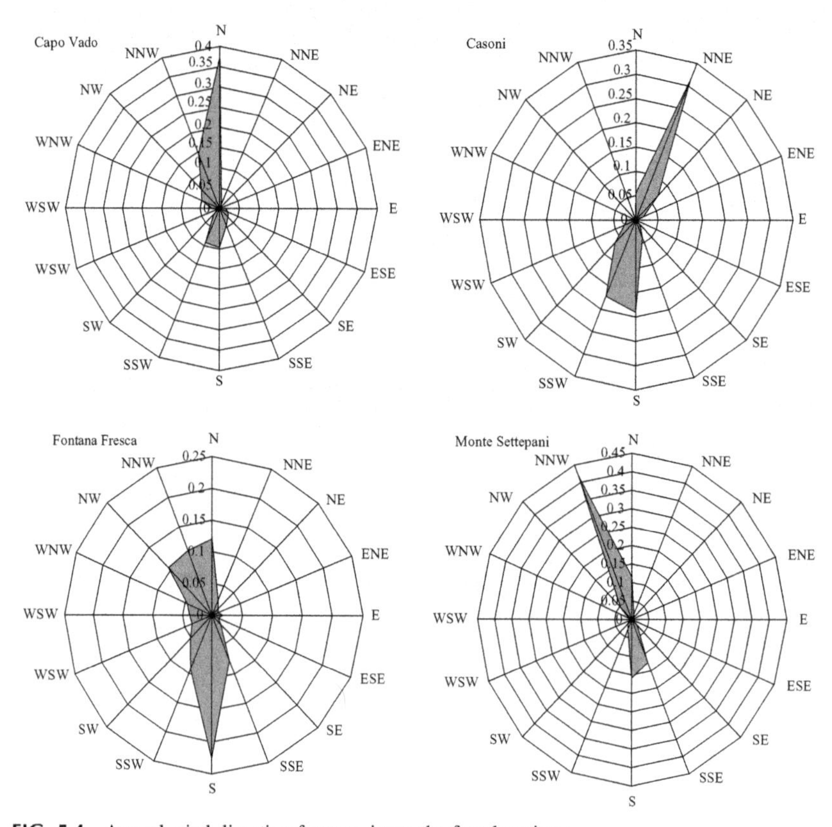

FIG. 5.4 Annual wind direction frequencies at the four locations.

6.94 m/s at 10 m. On the other hand, the wind speed reaches its minimum limit in July (4.69 m/s). At the hub height, the monthly wind speeds range between 6.33 and 9.36 m/s. The shape parameter k takes a value between 1.90 and 2.45, while the scale parameter c between 5.30 and 7.83 m/s.

To sum up, a maximum value of the mean wind speed at 10 m is obtained at Capo Vado as 9.98 m/s in December. Furthermore, the Weibull parameter k varies between 0.95 and 2.45 with a minimum at Capo Vado in April and a maximum at Monte Settepani also in April. The parameter c varies between 2.73 and 11.25 m/s, thus for a maximum and minimum values observed at Capo Vado in December and April, respectively.

In addition to the wind speed analysis, the knowledge of the wind direction is an essential task to carry out in order to make a better understanding regarding the planning of the wind turbine installations. The wind direction frequencies for the four locations are displayed in Fig. 5.4. The wind directions show a quiet similar behavior. But for Fontana Fresca, all the other locations (Capo Vado, Casoni and Monte Settepani) exhibit a distribution between NNE to NNW pattern. For Capo Vado, the predominant wind direction is the North with a value of 37%, while for Casoni and Monte Settepani, it is, respectively, NNE and NNW with 31% and 41%. An opposite tendency is observed at Fontana Fresca site, where the predominant direction is the South.

The estimation of the mean wind speed over a site is not a final step to assess the available wind potential in the considered site. Moreover, the value of the power density is an important parameter that can provide complementary information regarding the choice of suitable site, as well as an immediate classification of the site. For this main reason, the wind power density available at the four locations has been computed. Fig. 5.5 reveals the monthly variation of mean wind power density at different heights for the four selected stations. It can be recognized that all the four stations (Capo Vado, Casoni, Fontana Fresca, and Monte Settepani) exhibit the same tendency as regards the highest monthly mean wind power density and which occurs in December. Whereas, once it comes to the minimum wind power density, the rate of variation is not the same; the occurrence of the minimum differs from one site to the other. At 10-m elevation, this minimum is reached in August at Casoni site with a value of 108.32 W/m^2, and for Capo Vado in April with a value of 90.71 W/m^2, while for the two other sites Fontana Fresca and Monte Settepani, their minimum values occur in July with values equal, respectively, to 66.13 and 106.21 W/m^2. Comparing the trend of the four wind power density in Fig. 5.5, the wind power density has its maximum value in December for Capo Vado with 1177.97 W/m^2 at 10 m and 2411 W/m^2 at the hub height (76 m).

The seasonal wind characteristics for all the stations are presented in Table 5.8. It is observed that the highest value of the mean wind speed and the mean wind power density for all locations is observed in the Autumn season which coincides with the increased demand of energy. It is also apparent from the same table that Capo Vado location is the windy site, at which the maximum seasonal mean wind speed is about 8.76 m/s (at height of 10 m) observed in the Autumn season, while the minimum of 4.06 m/s in Spring, whereas the seasonal mean power density varies between 134.64 and 872.87 W/m^2 reached, respectively, in Spring and Autumn seasons. At the hub height (76 m), the seasonal mean wind speed occurred between 5.48 and 11.81 m/s. The obtained results for the other locations are reported in detail in Table 5.8.

The seasonal variation of the wind speed can help in forecasting the future trend of wind projects. Due to the random behavior of the wind speed and also its variation over time, it is more practical to represent its behavior using a probability function. The comparison between the actual seasonal data and the estimated seasonal Weibull frequency distributions of wind speed of the four locations is shown in Fig. 5.6. It can be seen that the Weibull distribution demonstrates a good fit. Furthermore, it is observed for Capo Vado and Casoni, that the wind speed covers the large range of variation in Winter and Autumn seasons, and which reach 0–20 m/s, whereas in Spring and Summer the higher range limit does not exceed 15 m/s. For Fontana Fresca and Monte Settepani, the

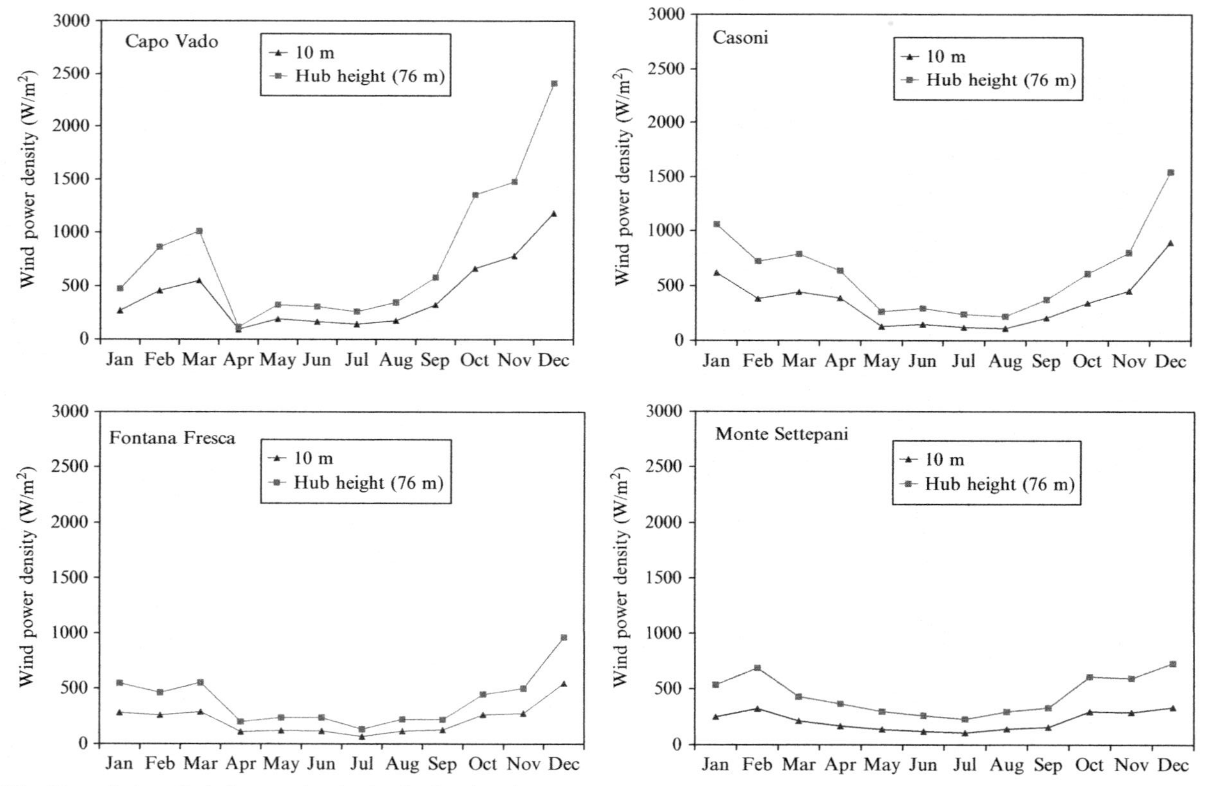

FIG. 5.5 Monthly variation of wind power density for the four locations.

TABLE 5.8 The seasonal wind characteristics

Season	c (m/s)	*k*	Mean wind speed (m/s) 10 (m)	Mean wind speed (m/s) 76 (m)	Power density (W/m^2)
Capo Vado					
Winter	7.09	1.51	6.39	8.63	422.47
Spring	4.41	1.32	4.06	5.48	134.64
Summer	5.71	1.56	5.13	6.92	208.45
Autumn	9.84	1.78	8.76	11.81	872.87
Casoni					
Winter	7.23	1.46	6.55	8.84	477.06
Spring	5.73	1.58	5.15	6.94	207.00
Summer	5.24	1.69	4.67	6.30	140.88
Autumn	7.36	1.39	6.71	9.06	555.13
Fontana Fresca					
Winter	6.40	1.62	5.73	7.73	274.30
Spring	4.93	1.74	4.39	5.93	113.15
Summer	4.51	1.58	4.05	5.46	100.46
Autumn	6.39	1.40	5.83	7.86	359.11
Monte Settepani					
Winter	6.98	2.08	6.18	8.34	260.94
Spring	5.90	2.39	5.23	7.06	140.25
Summer	5.63	2.12	4.98	6.72	134.16
Autumn	7.35	2.06	6.51	8.78	307.17

wind speed covers the large range of variation in Winter and Autumn which is equal to 0–15 m/s. In Spring and Summer the higher range limit does not exceed 10 m/s. Results of wind availability in Capo Vado site show that the wind speed is above 3 m/s, respectively, in Autumn, Winter, Spring, and Summer with 82%, 68%, 49%, and 63% of the time, so the wind power plant can produce energy for 82%, 68%, 49%, and 63% of the times, respectively, in Autumn, Winter, Spring, and Summer. The higher percentage of wind availability in Winter occurs at Casoni and Fontana Fresca, respectively, with 70% and 67%, and at Monte Settepani in Autumn with 79%.

The histograms of the monthly variation of the mean wind energy produced by the WPP and the available wind energy in the swept rotor area for the four locations are shown in Fig. 5.7. It is important to underline that, for the whole four locations, the highest energy produced by WPP can be reached in December, respectively, with 3800, 2439, 1519, 1146 MWh. Further seasonal assessments of the available and produced energy are reported in Fig. 5.8. For all sites, the produced energy by the WPP shows a large variation from season to season, in addition, the highest production values for all stations

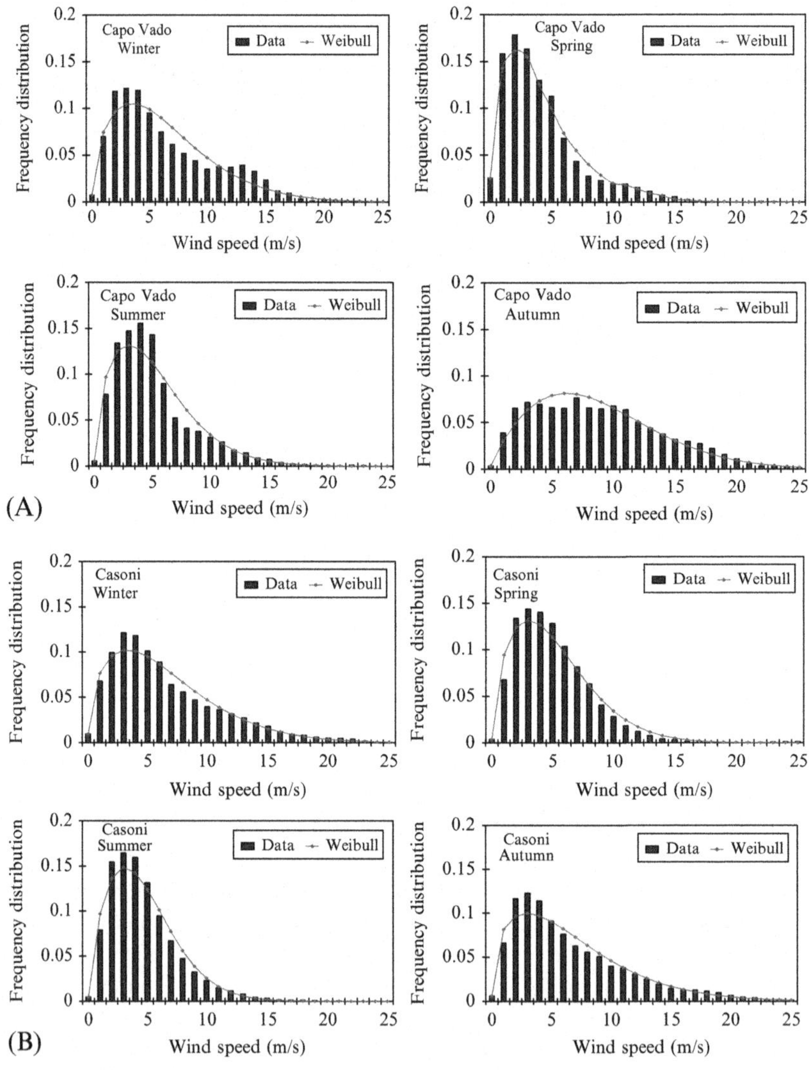

FIG. 5.6 Seasonal histograms of the wind speed. (A) Capo Vado, (B) Casoni,

Continued

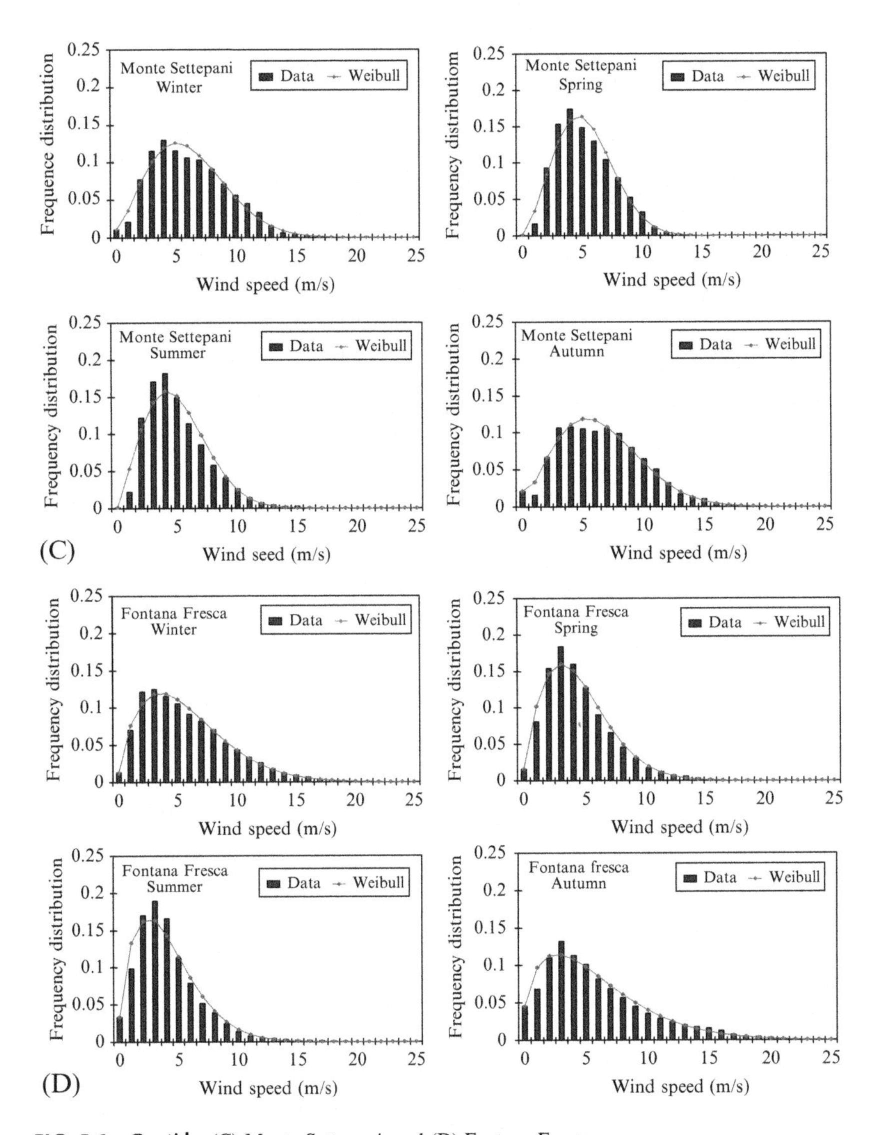

FIG. 5.6—Cont'd (C) Monte Settepani, and (D) Fontana Fresca.

occur in Autumn season with 8164, 4544, 2951, and 3039 MWh, respectively, at Capo Vado, Casoni, Fontana Fresca, and Monte Settepani.

The monthly and seasonal wind data analysis has been carried out to investigate wind characteristics and WPP production during the periods of 2002–08, 2006–08, and 2004–08 for, respectively, Casoni, Capo Vado, Monte Settepani, and Fontana Fresca. The monthly and seasonal wind speed distribution, wind power densities, and wind direction are determined for the four locations in

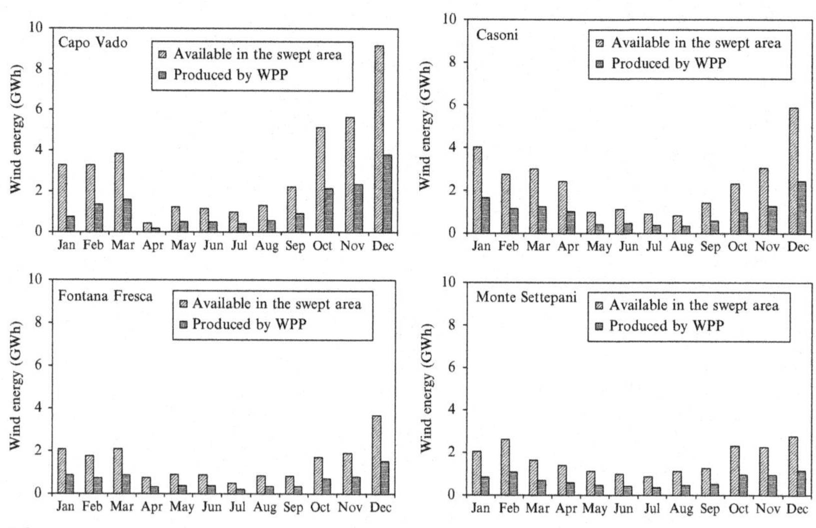

FIG. 5.7 Histograms of the monthly of the mean wind energy produced by the wind power plant versus the available wind energy in the swept rotor area for the four locations.

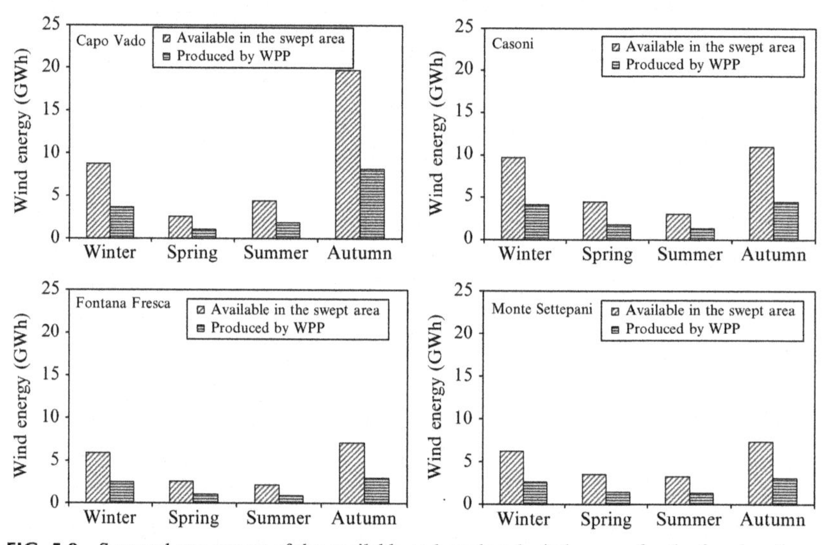

FIG. 5.8 Seasonal assessment of the available and produced wind energy for the four locations.

order to provide information of wind resources; further assessment of the monthly and seasonal wind energy available in each site and the energy output of the WPP have been done. It is believed that Capo Vado is the best site with a monthly mean wind speed determined between 2.80 and 9.98 m/s in December at a height of 10 m and a monthly wind power density between 90.71 and 1177.97 W/m^2 while the highest energy produced by WPP was reached in December with 3800 MWh.

As a result of the Battelle-PNL classification made in previous section—Capo Vado Class 7, Casoni Class 6, Fontana Fresca and Monte Settepani Class 4—all the four sites are considered to be suitable for most wind turbine applications taking into account data on the whole year. On the other hand, for example, Capo Vado dramatically falls in Class 1 if data limited to the month of April are taken into account and in Class 7 if limited to the months February, March, October, November, and December. This fact should reflect the inadequacy of Battelle-PNL classification on regions with a complex orography and variable wind characteristics as Liguria region, and, in general, as many others Mediterranean countries.

The seasonal variation of these sites should reflect the need of adopting proper strategies to adapt the wind exploitable energy to the demand. These could be done according to two main—not alternative—strategies: hybridizing the production with the contribution of some other renewable energies; storing energy to feed future demand with the aim to avoid shortage. As regards the former option, for the sites investigated in this work, solar energy should be a promising option, since sun irradiation reaches high values just in the months and seasons where wind seems to have lower energy exploitation. As regards the latter option, hydrogen might be a challenging way to store energy, specifically if coupled with automotive hydrogen fuel future demand. For example, as a rough estimation, for the site of Capo Vado, the estimated hydrogen gas mass that can be produced using an electrolyzer characterized by an efficiency of 0.9 is about 65 tones in December (2166kg/day) which is equivalent to 2558MWh of hydrogen energy production and 3 tones in April which is equivalent to 117MWh of hydrogen energy production. In this respect, proper strategies should be studied to couple the storage of hydrogen for household and industrial energy consumption and as a fuel for transport vehicles. From an automotive perspective, as a kilogram of hydrogen is roughly equivalent to a gallon of gasoline in energy content, assuming a 6kg/fill average per day for a car, the 2166kg/day would fill approximately 360 cars per day. For a hydrogen vehicle with internal combustion engine fuel consumption is about 0.60kWh/km as reported by (Ajanovic, 2008), thus the hydrogen energy production in December for Capo Vado site is more than 4×10^6km. Table 5.9 present the solar energy available as computed by statistical analysis and the geographical coordinates of the 16 meteorological stations. It can be seen that the obtained solar energy shows quiet similar values which is mainly due to the similar orography and climate. The results presented in Table 5.9 have been used to map the solar potential of Liguria region using an Arcgis tool.

The theoretical solar potential of Liguria region that can be exploited for energy production is reported in Fig. 5.9. As it can be seen, the total annual solar energy is ranged between 1128 and 1534kWh/m^{2}yr, which are reached, respectively, in Buorgonuovo (code 6) and Sanremo (code 11). In fact, the estimated solar energy over the region does not present important differences between all locations of the territory.

TABLE 5.9 Solar energy production

No.	Sites	E_{solar} (kWh/ m^2yr)	Latitude N	Longitude E	Altitude (m)
1	Castellari	1239.79	44.1456	8.2625	100
2	Genova	1384.34	44.4017	8.9472	20
3	Polanesi	1356.08	44.3658	9.1247	50
4	Monterocchetta	1497.05	44.0755	9.9197	412
5	Corniolo	1512	44.1063	9.7348	338
6	Buorgonuovo	1144.6	43.8463	7.6208	100
7	Capo Vado	1456.27	44.2583	8.4425	170
8	Cavi di Lavagna	1365.75	44.2961	9.3739	100
9	Levanto\ S.Gottardo	1421.11	44.1811	9.6211	100
10	Allassio	1449.24	44	8.17	10
11	Sanremo	1556.32	43.82	7.78	45
12	Romito Magra	1345.26	44.1033	9.9303	100
13	Ranzo	1380.08	44.0632	8.0049	310
14	Pornassio	1273.39	44.0639	7.8664	500
15	Giacopiane Lago	1375.59	44.4608	9.3875	1016
16	Poggio Fearza	1331.9	44.042	7.7935	1833

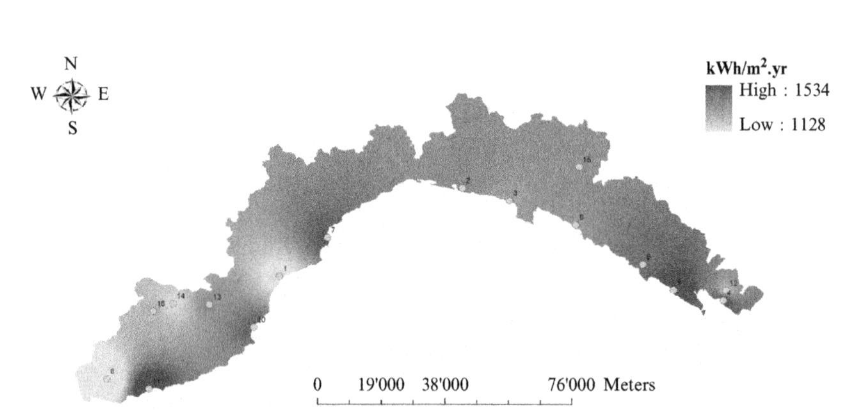

FIG. 5.9 Solar energy production map.

3. SELECTION CRITERIA FOR ALTERNATIVE TRANSITION TOWARD THE IMPLEMENTATION OF HYDROGEN INFRASTRUCTURE

3.1 Introduction

Many studies have claimed that refueling infrastructures are one of the most formidable barriers to a successful transition to a hydrogen-based road transportation system (Melendez, 2006). To the author's knowledge, to date, papers have treated the problem from the viewpoint of hydrogen demand, costs, market penetration, and distance to fuel hydrogen vehicles and areas of interest. More emphasis must be given upstream the hydrogen refueling stations network, in particular to criteria of selection related to the locations of resources of production. The overall objective is to present an integrated approach considered as a decision support system (DSS) for the selection of hydrogen refueling stations. The method combines two different approaches: a detailed spatial data analysis using a geographic information system with a mathematical optimization model of set covering. Regarding GIS component, criteria related to the demand and safety are considered for the selection of the hydrogen stations, while in the mathematical model, criteria that regard costs minimization are considered. In general, the DSS will identify the suitable sites providing information on multicriterion level evaluation of locating the hydrogen infrastructure. The GIS component of the integrated approach will be applied to the region of Liguria, North of Italy, while the application of mathematical model is projected as a future work since the model is still in the development phase.

3.2 Method Description

3.2.1 GIS-Based Approach Decision

The GIS-based modeling is implemented as a first stage of the DSS in order to find whether a location is convenient to be a future hydrogen refueling station or not.

The geographic information system is employed to study the spatial relationships that exist between the hydrogen demand, future location, primary resources, and existing petrol infrastructure that could be considered as a potential site for hydrogen exploitation. This tool requires the use and integration into one component sources of spatial data from national organizations, local authorities, or private companies for the characterization of the area. It helps to snow the nature and location issues such as population, personal property, public buildings, water system, transport infrastructure, areas of nature conservation, etc. (Garbolino, n.d.).

This section of GIS-based modeling is developed at regional scale considering various market hydrogen penetration levels. The analysis was done based on previous works of Johnson et al. (2008).

Even though it is important to highlight that decision support systems based on GIS alone is not ideal for the planning of future infrastructure, in particular, hydrogen refueling stations. In a problem location, the decision makers usually need to perform a selection study, and to choose among a combination of a large set of candidate sites which satisfy some predefined criteria. The GIS module can be considered as a preliminary step to select the eligible hydrogen sites. As a result, the GIS tool could lead to a significant reduction of the number of locations that need to be implemented in the mathematical optimization problem, which may reduce significantly the computational time.

3.2.2 Mathematical Optimization: Model Description

The second component of the approach deals with the mathematical optimization, where eligible hydrogen refueling stations locations, found by the GIS component are implemented as an input to find the optimal set of hydrogen refueling stations that satisfy optimization objectives.

The mathematical optimization is cost driven where the main objectives are to minimize the cost of installation of new on-site hydrogen refueling stations, the cost of conversion of existing gasoline to hydrogen stations, and the cost of transporting hydrogen fuel to off-site stations. The model takes into account the cost minimization in order to select those hydrogen refueling stations, a priori—previously selected—by the first model based on the geographical information system.

The novelty of such study is to develop a decision support system for the localization of hydrogen refueling stations taking into account the potential of production within a specific boundary region. Fig. 5.10 represents the general model components and architecture.

3.2.3 Model Characteristics

3.2.3.1 Hydrogen Refueling Stations: Type and Scale

It is important to understand the function of the service station, which may simply be a dispensing station (off-site hydrogen refueling station), having hydrogen stored in reservoir or both a production and dispensing station (on-site hydrogen refueling station). Fig. 5.11 displays different configurations of renewable hydrogen refueling stations that may exist.

One particularity of the proposed approach is the adoption of a combination of on-site and off-site hydrogen refueling stations that are considered as an input for the model.

During the current study, two configurations are considered:

- The on-site hydrogen refueling stations (Fig. 5.12)
- The off-site hydrogen refueling stations (Fig. 5.13)

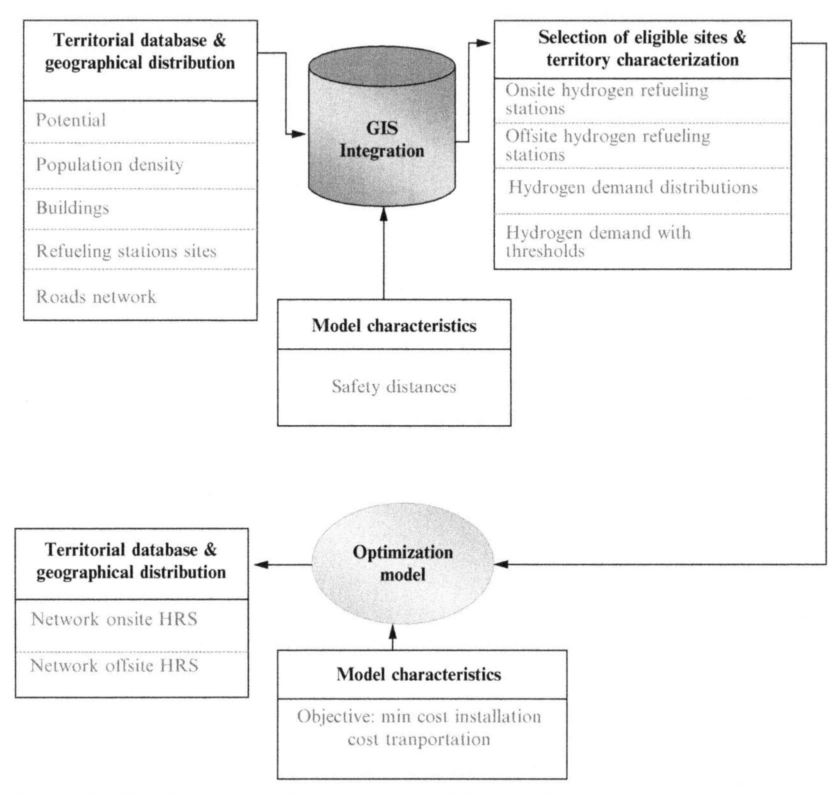

FIG. 5.10 Model components displaying both modules considered.

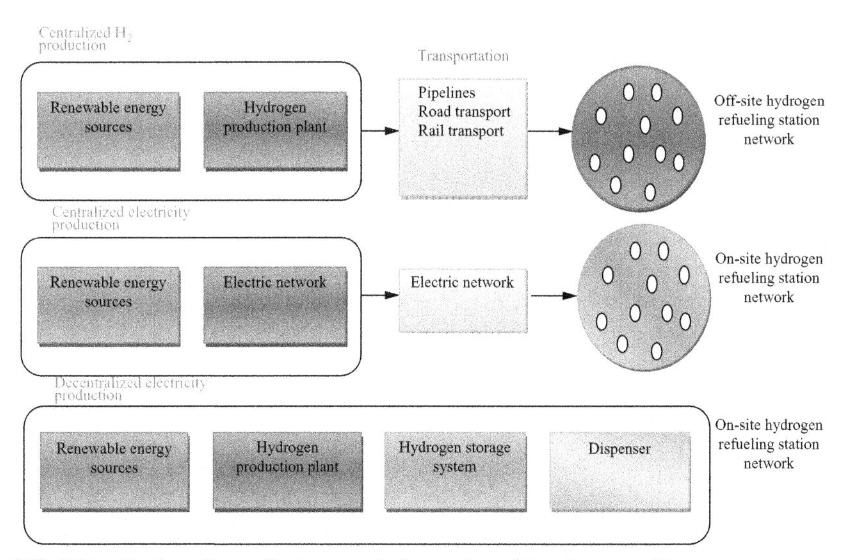

FIG. 5.11 Configurations of hydrogen refueling stations driven by renewable energy.

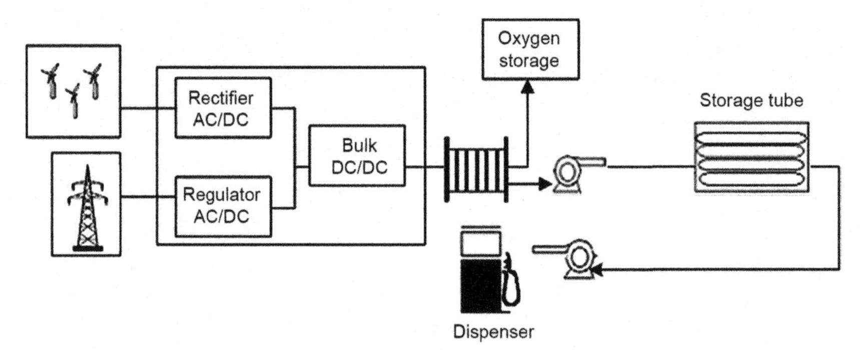

FIG. 5.12 Electrolysis on-site hydrogen refueling station.

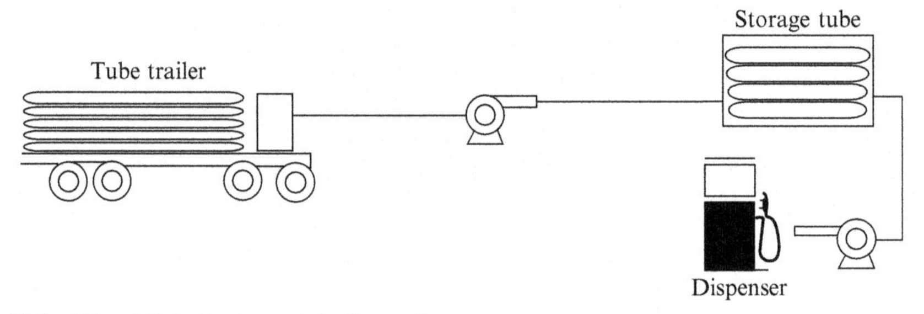

FIG. 5.13 Off-site hydrogen refueling station.

However, the main goal is to determine the optimal placement for the two types of hydrogen refueling stations noted previously. The problem has been solved assuming two scenarios for hydrogen refueling stations:

- Scenario 1: conversion of petrol stations to hydrogen stations. This scenario is been implemented assuming that petrol companies will then represent one of the major categories of hydrogen producer. This analysis can also be undertaken to avoid somehow the most haphazard placement of hydrogen station.
- Scenario 2: installation of new hydrogen stations. The scenario has been implemented to take into account the on-site hydrogen station, because a priori scenario 1 alone may not cover the demand due to the geographical location and hydrogen production potential within the station.

3.2.3.2 GIS Selection Criteria

- Hydrogen demand

The hydrogen demand is one of the important criteria that must be taken into account for the localization of the station. From a demand viewpoint, the selection will depend on the spatial distribution of the hydrogen demand based on the specific market and population data.

Methodology proposed by Ni et al. (2005) is adopted and adapted according the proposed case study. The calculation method is based on the population, a multiplier of the number of vehicles per person, another multiplier for the fraction of hydrogen that use hydrogen fuel, and a multiplier related to fuel economy of hydrogen vehicles as given in the following equation:

$$\text{Hydrogen demand density}\left[\text{kg}\,H_2/\text{day}/\text{km}^2\right] = \text{population density}\left[\text{people}/\text{km}^2\right]$$
$$\times \text{vehicle ownership}\,[0.7\,\text{LDV}/\text{person}] \times \text{market penetration rate}\,(5\%, 10\%,$$
$$40\%, 60\%, 80\%, \text{and } 100\%) \times \text{fuel use}\,(0.6\,\text{kg}\,H_2/\text{day}/\text{vehicle})$$

Based on the GIS, the hydrogen demand is calculated on a spatial basis, where various threshold values are used to identify demand density areas, since only those areas with sufficient hydrogen demand can be assumed to be viable locations for refueling stations opening.

It is important to underline that the hydrogen demand also depends strongly on the market penetration of the hydrogen vehicles. However, the accurate knowledge of the hydrogen market penetration is a hard task, especially because of the uncertainty in supply and demand (related to the chicken and eggs dilemma). Consequently, the current model adopts steady scenario assuming various values for the hydrogen market penetration. Six scenarios for hydrogen market were implemented here (6%, 10%, 40%, 60%, 80%, and 100%). For instance, at 6% market penetration, it is assumed that 6% of the vehicles in the entire region are hydrogen fuel cell vehicles that are in operation within these areas.

- Safety criteria

Another criterion that may play increasing role in the selection process of hydrogen refueling stations is related to the safety issues. In fact, refueling station must be adequately located in order to better serve demand of a specific category of hydrogen fuel vehicle users, but, in the same time, they must avoid high risk that can bring to populations and environment.

Two methods could be used to evaluate whether a location can be considered as eligible one or not.

These methods are risk assessment (quantitative and qualitative) and analysis of safety distances. The latter is adopted to justify the permitting process of hydrogen refueling station. The choice of such approach is related to lack of complete data to complete the quantitative risk assessment. The current work will estimate the safety distance from a refueling station based on an approach by consequence, where consequence of worst possible scenario is considered to set adequate safety distances. In this context, explosion represents the worst scenario; as stated by Zhiyong et al. (2010), the explosion produces the longest harm effect distances, both to people and to equipment. The main attention to the introduction of hydrogen refueling station presented will be a macro-scale level, which reflects the outside risk to people and environment nearby.

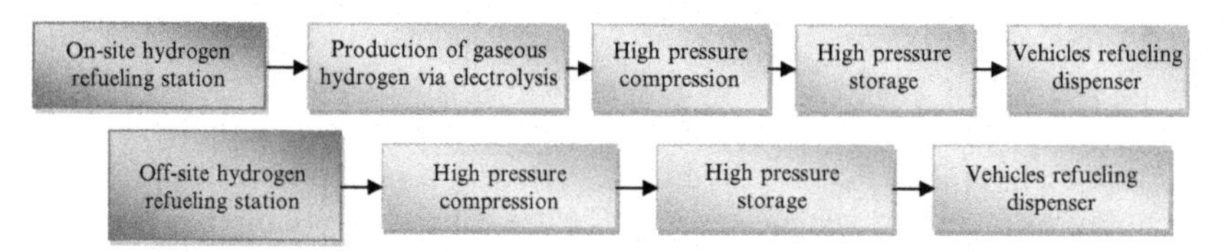

FIG. 5.14 General components of hydrogen refueling station.

This method is recognized by many authors to be an effective method, which must be used primary in the approval process of hydrogen refueling stations.

3.3 Hydrogen Refueling Station Components

The safety distance from a hydrogen refueling station will be determined assuming gaseous fuel stations. These station types may include equipment for the supply, compression, storage, and dispensing of fuel. The failure may occur in any part leading to a scenario of high consequence. The distance required to mitigate the exposure of the target would differ depending on what is the source of the release or kind of failure. Fig. 5.14 presents the general components of on-site and off-site HRS that will be considered.

Hydrogen production system: hydrogen is produced on-site by electrolysis process driven by wind turbine of electricity coming from different sources, namely, wind, solar, and electric network. The electrolysis process is driven by an alkaline type, which is the most mature electrolysis technology used today. It is assumed that one unit electrolyzer is installed. It operates at a rate of 10 Nm^3/h, with a consumption of water equal to a value of 7.8 lt/h. Hydrogen is produced at a pressure of 4 bar. Hydrogen compressor system: the compressor is an essential element of the refueling station, whenever the type of this latter (on-site or off-site). Here, the compressor allows a compression of hydrogen from 4 to 200 bar. The compressor is installed with the electrolyzer in the same enclosure. Hydrogen storage system: The hydrogen station is equipped with hydrogen cylinders for the storage of hydrogen, where the capacity of each is between 30 and 300 kg. The main difference between the stations' storage capacities lead into their capacities to refueling cars (how many cars are refueled daily) and the quantity of hydrogen storage in the inventory of each one.

3.4 Consequence of Worst Scenario: Explosion

- Scenario of failure at the hydrogen refueling station

The hazard source within a hydrogen refueling station can be any object (installation, equipment, construction, machinery, etc.) that handle hydrogen and which may create a hazard to its surroundings. Gaseous hydrogen refueling

TABLE 5.10 Accidents leading to explosion for on-site and off-site station (CEC: California Energy Commission, 2005; http://www.h2incidents.org)

Components involved	Scenarios	Failure
Off-site hydrogen refueling station	Tube trailer	Accident, compressed trailer leak, failure of pressure relief device, failure of tube on trailer
	Compressor	Valve on discharge of compressor fails closed, high pressure hydrogen supply, line to station failure, leak from compressor
	Storage tank	Relief device failure (on cylinders) fails open, storage tank failure, piping leak
	Dispenser	Piping failure, drive away while connected to dispenser, hose failure, vehicle pressure relief device leaks, vehicle tank backflows through, dispenser vent
On-site hydrogen refueling station	Electrolyzer system	Exposed electrical circuit, rectifier startup, hydrogen gas leak or full rupture of pipe, electrolysis gas vent valve leaks
	Compressor	Compressor line failure, compressor seal failure, failure compressor valve, hydrogen supply line failure
	Storage tank	Overpressure and fail storage tank, storage tank failure, relief device failure fails open
	Dispenser	Piping failure, drive away while connected to dispenser, hose failure, vehicle pressure relief device leaks, vehicle tank backflows through dispenser vent

stations either on-site or off-site mainly consist of compressor, storage tank, piping system which link different components to each other. Owing to this architecture, the hazard may arise from component failure, operational mistake from various sources, and in a number of ways (leak, full bore, etc.). The distance required to mitigate the exposure of the target would differ depending on what is the source of the release or kind of failure. Table 5.10 reports different accidents that may derive from different components of the stations which may result of potential explosion of gaseous hydrogen, two databases have been used (CEC: California Energy Commission, 2005). These accidents can lead to a potential of exposing the third party (outside the station), which are of our interest in this study:

3.5 Harm Criteria for Explosion

3.5.1 Harm to People

The possible effect of overpressure on humans includes direct and indirect effects. The main direct effect is the sudden increase in pressure that can cause damage to pressure-sensitive organs such as the lungs and ears. Indirect effects include the impact from fragments and debris generated by the overpressure event, collapse of structure, and heat radiation (from fireball generated by vapor cloud explosion) (Jeffries et al., 1997). For people, harm criteria can be expressed in terms of death or injury. The level of person harm is dependent upon factors such as the age of the person and the exposure time. Table 5.11 represents the damage caused by an overpressure to people.

It can be shown from Table 5.11 that the threshold overpressure for no "direct" harm is 0.138 bar, while no "indirect" harm is equal to an overpressure of 0.069 bar.

A mathematical parameter that can join the people harm caused by the overpressure to the peak overpressure is named probit function. The probit function

TABLE 5.11 Direct and indirect effect on people damage due to overpressure (Jeffries et al., 1997)

Effect type	Overpressure	Damage description
Direct effect	0.138	Threshold for eardrum rupture
	0.345–0.483	50% probability of eardrum rupture
	0.689–1.03	90% probability of eardrum rupture
	0.82–1.03	Threshold for lung hemorrhage
	1.38–1.72	50% probability of fatality from lung hemorrhage
	2.07–2.41	90% probability of fatality from lung hemorrhage
	0.48	Threshold of internal injuries by blast
	4.83–13.8	Immediate blast fatalities
Indirect effect	0.10–0.20	People knocked down by pressure wave
	0.14	Possible fatality by being projected against obstacles
	0.55–1.10	People standing up will be thrown a distance
	0.069–0.13.8	Threshold of skin lacerations by missiles
	0.28–0.34	50% probability of fatality from missile wounds
	0.48–0.69	100% probability of fatality from missile wounds

that estimates the fatality level for explosion due to overpressure is expressed in percentage and is given by (Eisenberg et al., 1975):

$$Pr = -77.1 + 6.91\ \ln(P_s)$$

where P_s is the overpressure.

3.5.2 Harm to Structures

According to Table 5.12, the overpressure caused by the explosion has a minimum threshold equal to 0.01 bar and which correspond to the threshold for glass breakage.

Similarly to harm to people, Eisenberg et al. (1975) have introduced a mathematical formula that enables the calculation of the probit function due to harm to structure and other equipment. The formula calculates the total damage that results from an overpressure.

$$Pr = -23.8 + 2.92\ \ln(P_s)$$

where P_s is the overpressure.

An estimation done by Zhiyong et al. (2011) has assumed that 100% lethality is obtained for an overpressure of 0.3 bar and 0% lethality for lower overpressure levels for people outdoors (outside boundary of the station). Results published by Rosyid (2006) are used, which conclude that the effect zone due to an explosion could be represented by a circle or ellipse centered at the release point of hydrogen. In their results, the effects zone due to the explosion caused by a liquid release is much more higher than those reached for the case of gaseous hydrogen release by about three times. This remark could enhance the safety by using gaseous HRS.

TABLE 5.12 Damage to structures and equipment due to overpressure (American Institute of Chemical Engineers, 1998)

Overpressure (bar)	Damage description
0.01	Threshold for glass breakage
0.15–0.20	Collapse of unreinforced concrete or cinder block walls
0.2–0.3	Collapse of industrial steel frame structure
0.35–0.4	Displacement of pipe bridge, breakage of piping
0.7	Total destruction of buildings; heavy machinery damaged
0.5–1	50e100 Displacement of cylindrical storage tank, failure of pipes

TABLE 5.13 Threshold between effect distances for "no harm"

Type	Station size	Storage capacity (kg)	Safety distance (m)
On-site Electrolysis hydrogen refueling station	Small	30	150
	Large	420	334.4 (early explosion) 341.2 (late explosion)
Off-site compressed hydrogen refueling station	Small	30	100
	Medium	100	125
	Large	300	300

Table 5.13 presents the safety distance assumed in this study, for the two types of hydrogen refueling stations, and for different capacities of inventories available at the station. It can be shown that the capacity of inventory has a major role to define the radius of the effect zone around the station.

3.6 GIS Based on Safety Distance Criteria

Through the definition of the harm distances or distance with high risks, GIS is deployed to create an overlapping map between the hydrogen refueling station to be installed on the specific territory and the harm distance observed in case of explosion around the station. This procedure is entitled hazard mapping. In our approach, once developing the hazard mapping, another overlapping will be created by combining this latter with the density of population of the region and land use of the region. Refueling stations situated in a territory with high population density and/or harm distance will not be considered for the conversion or installation (Fig. 5.15).

The GIS-based decision support system is performed. It consists of four modules that are described hereafter:

Module 1: Buffer layer: The aim is to draw a buffer around all stations implemented in the study. The dimension of the buffer will reflect the safety distance associated with each type of hydrogen refueling station to be implemented. The buffer regards the existing petrol stations that have the potential to be converted to hydrogen refueling station, and also the new on-site station that are set depending on the hydrogen production potential available within the territory.

Module 2: Population Intersect layer: During this phase, an intersection has been made between the buffers of all stations and the population of the region.

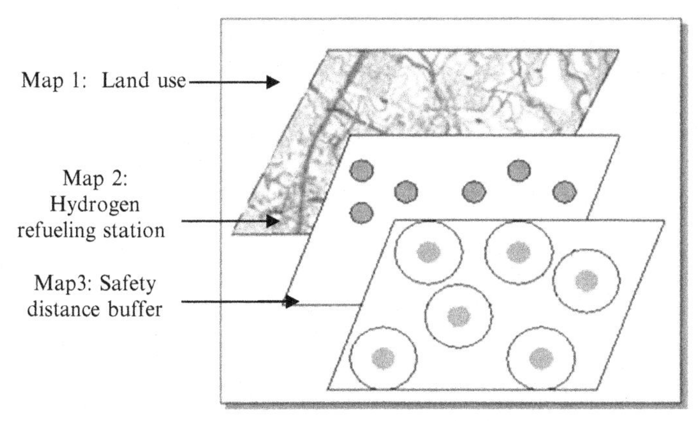

FIG. 5.15 GIS based study on risk-based decision support.

The aim was to examine the location of the stations according to different levels of population. Three levels were made regarding the population. When the population is up to 50 thousands, the region can be categorized as Level I. Level II corresponds to region with population between 5 and 50 thousands and Level III for population under 5 thousands.

Module 3: Buildings Intersect layer: the same analysis was made to assess the distribution of station according to the spread and levels of buildings available within the Liguria region. Similarly, three levels were made to classify the density of presence of buildings around the hydrogen station. Level I, buildings number up to 20 thousands; Level II for presence of buildings between 7000 and 50,000; and finally Level III with buildings number under 7000.

Module 4: This module aims to create a risk ranking matrix for both levels related to the population as well as the buildings. The aim of this module is not to make a detailed qualitative risk management, but just to exclude those stations that have a higher rank regarding population as well as buildings number. We claimed that station within Level I for both population and buildings number is excluded.

3.7 Case Study

The approach presented earlier is applied at a regional scale to a case study in the north of Italy. Various data are gathered and presented such as the availability of primary energy sources and their distribution, the hydrogen demand over the planning horizon, and the future possible scenarios of hydrogen infrastructure. In this case study, hydrogen is assumed to be produced from renewable-based electricity generation with the possible combination with the electrical network. The "clean feedstocks" in terms of renewable energy resources, mainly driven by solar and wind energy.

3.7.1 Hydrogen Demand Data

A GIS-based method to model the magnitude and the spatial distribution of hydrogen demand is developed. The data used to perform the estimation of hydrogen demand maps is mainly articulated around the population data of the region, the area of the region to be studied, beside some other technical parameters. Steps used are described hereafter:

1. The population density of the region (people/km^2).
2. An estimation of the hydrogen market penetration (%)
3. An estimation of the total vehicle available: (vehicles/km^2) vehicle (or auto) ownership multiplied by the population density. For Italy, the ownership is assumed to be equal to 0.571 vehicles/persons.
4. An estimation of the hydrogen vehicle density (H_2 vehicles/km^2): obtained by multiplying the market penetration by the total vehicle available
5. An estimation of hydrogen demand density (H_2 kg/km^2): obtained by multiplying the hydrogen vehicle density by an average vehicle fuel use of 220 H_2/year/vehicle. This last amount is estimated based on the fact that an average vehicle travels 25,000 km/year and has a fuel economy of 105 km/kg (equal to the one of gasoline by gallon)

Fig. 5.16 displays the percentage of hydrogen demand covered by the main populated areas (Genoa, Savona, La Spezia, Albenga, Ventimiglia, Sanremo, Imperia, Chiavari, and Rappallo). The histograms show the hydrogen demand in each area as a percentage of the total hydrogen demand in the region, for the specific scenario. It can be seen that Genoa city needs around 54% of the whole demand of the region. Maps of hydrogen demand for the six different scenarios are given in Fig. 5.17. It can be shown that hydrogen demand will begin with

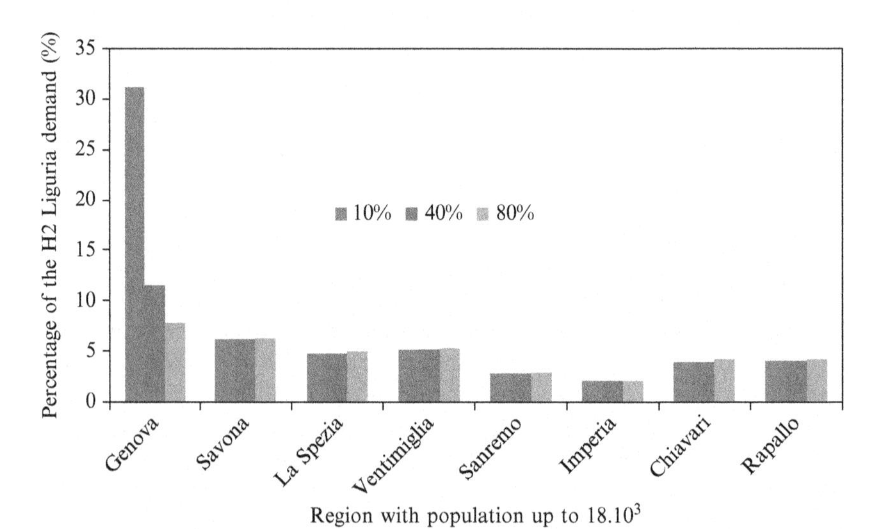

FIG. 5.16 Percentage of hydrogen demand in most populated areas in Liguria.

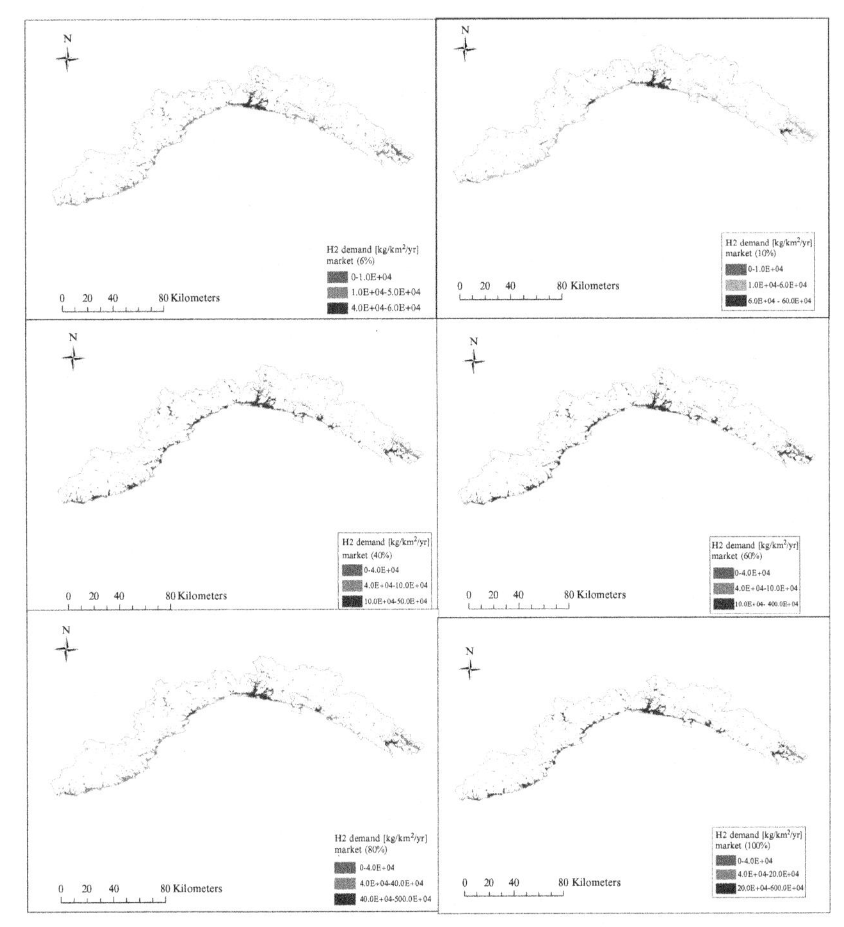

FIG. 5.17 Snapshot of hydrogen demand for different scenarios.

area with high population (up to 20,000) for small hydrogen market scenario (6, 10%) with a value between 60,000 and 600,000 kg/year. This remark seems obvious since population density centers will be the ones that will implement the hydrogen fuel cell vehicles.

3.7.2 *Locations of Hydrogen Refueling Stations*

The geographical location of the future hydrogen petrol station is based on the analysis of this information related to hydrogen potential production and main roads of the region. The most likely sites for alternative fuel refueling infrastructure were identified to satisfy the hydrogen demand and to minimize the risk. Different criteria were used to select several potential locations for hydrogen refueling stations including available infrastructure (roadways and highways), hydrogen market penetration, accessibility (remoteness of the station is avoided), locations of promising potential alternative fuel users, demand, and population. Information related to existing petrol stations and locations available for use as refueling sites were gathered and displayed in Fig. 5.18. While Fig. 5.19 displays the potential sites for the on-site hydrogen production.

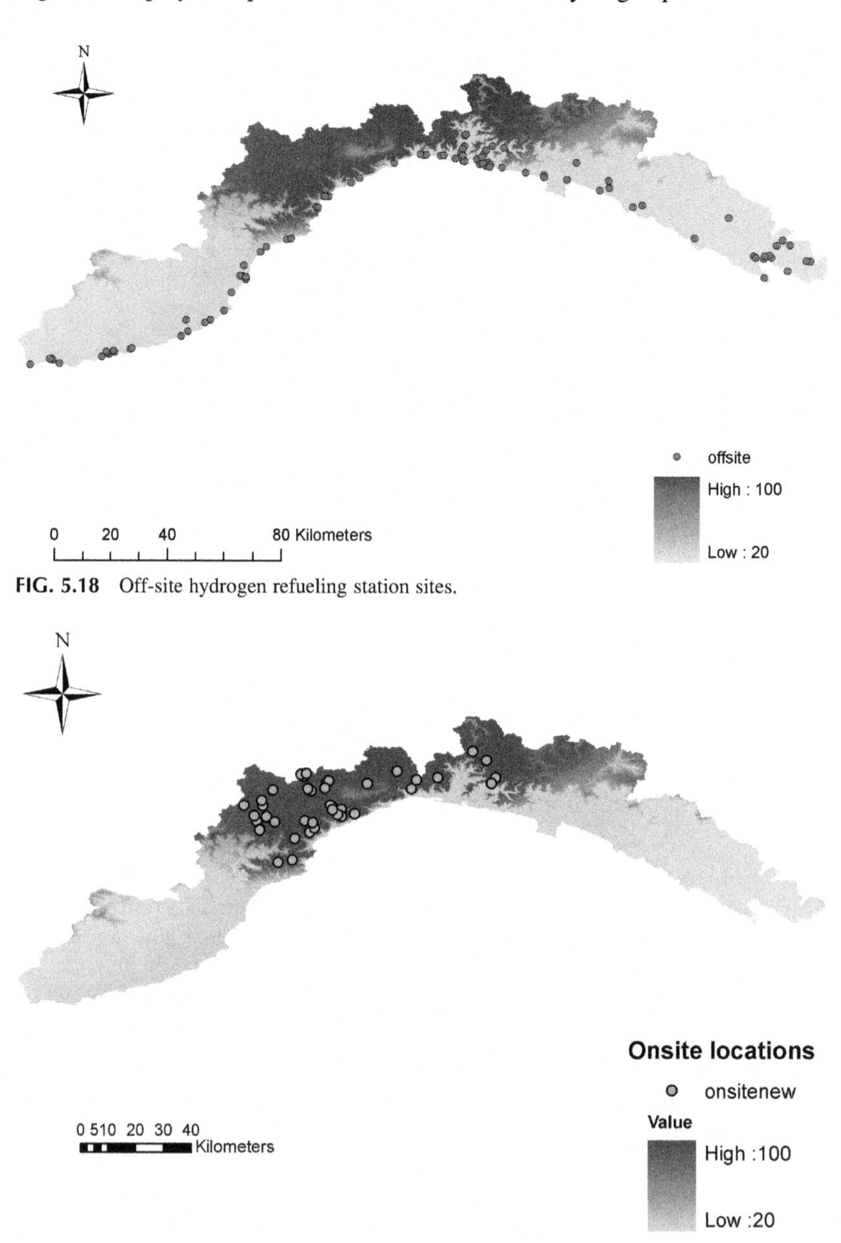

FIG. 5.18 Off-site hydrogen refueling station sites.

FIG. 5.19 On-site hydrogen sites.

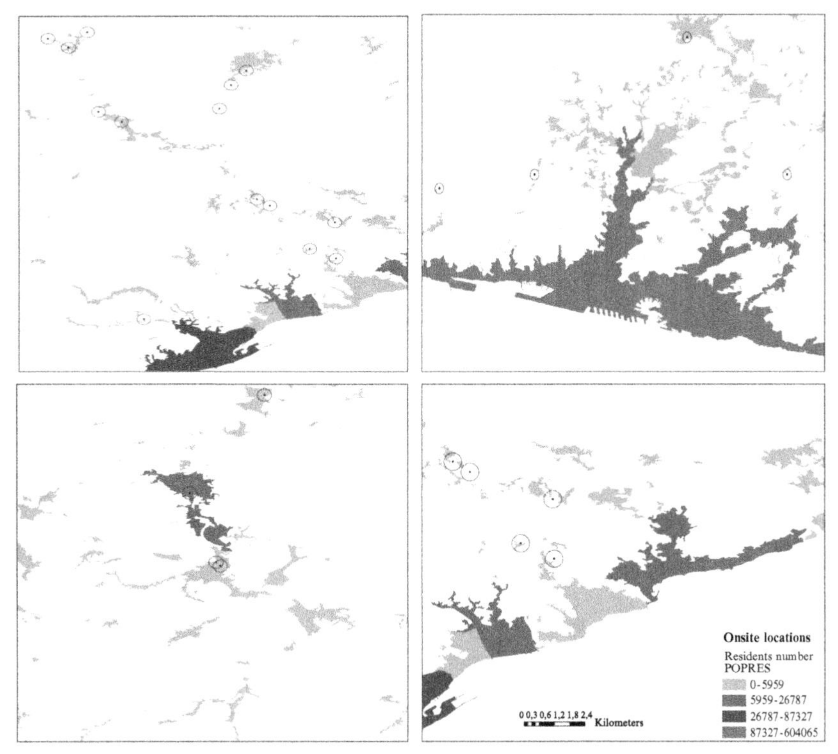

FIG. 5.20 On-site buffer location versus population.

3.7.3 Eligible Hydrogen Refueling Stations

As mentioned previously, the decision support related to the risk aims to exclude all stations that can bring higher risk to population and environment. Fig. 5.20 displays the buffer of the on-site location, while Fig. 5.21 is related to the buffer of the off-site locations. This GIS-based decision support system, even it is a rough one, but it has enabled the reduction of 28 off-site stations from a number of 93. As regards the on-site station, the procedure has enabled the exclusion of two stations. This remark is due to the low density of hydrogen in places of high hydrogen production potential. This remark could favor the installation process of new on-site station operating on the renewable energy resources.

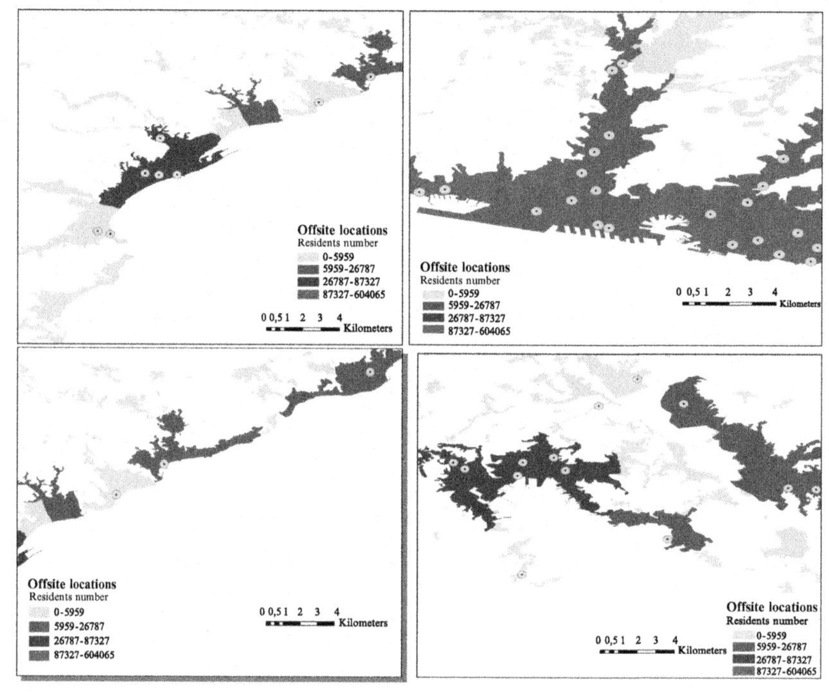

FIG. 5.21 Off-site locations versus population.

REFERENCES

Ajanovic, A., 2008. On the economics of hydrogen from renewable energy sources as an alternative fuel in transport sector in Austria. Int. J. Hydrog. Energy 33, 4223–4334.

American Institute of Chemical Engineers, 1998. Guidelines for the evaluation of the characteristics of vapor cloud explosions, flash fires, and bleves, center for chemical process safety. American Institute of Chemical Engineers.

Carta, J.A., Ramırez, P., Velazquez, S., 2009. A review of wind speed probability distributions used in wind energy analysis. Case studies in the Canary Islands. Renew. Sust. Energ. Rev. 13, 933–955.

CEC: California Energy Commission, 2005. Failure modes and effects analysis for hydrogen fueling options. CEC-600-2005-001.

Celik, A.N., 2006. A simplified model for estimating yearly wind fraction in hybrid-wind energy systems. Renew. Energy 31, 105–118.

Conte, A., Pavone, A., Ratto, C.F., 1998. Numerical evaluation of the wind energy resource of Liguria. J. Wind Eng. Ind. Aerodyn. 74, 355–364.

Eisenberg, N.A., Lynch, C.J., Breeding, R.J., 1975. Vulnerability model: a simulation system for assessing damage resulting from marine spills. Final Report SA/A-015 245, US Coast Guard.

Garbolino E. Contributions à l'étude du risque de transport sur route de marchandises dangereuses dans un espace transfrontalier. Habilitation à Diriger des Recherches en Géographie. HDR thesis. Université de Nice—Sophia Antipolis.

Himri, Y., Stambouli, A.B., Draoui, B., 2009. Prospects of wind farm development in Algeria. Desalination 23, 130–138.

IEA energy technology essentials. OECD/IEA. http://www.iea.org/techno/essentials6.pdf.

Ilinca, A., McCarthy, E., Chaumel, J.L., Rétiveau, J.L., 2003. Wind potential assessment of Quebec Province. Renew. Energy 28 (12), 1881–1897.

Jeffries, R.M., Hunt, S.J., Gould, L., 1997. Derivation of fatality of probability function for occupant buildings subject to blast loads. Health & Safety Executive.

Johnson, N., Yang, C., Ogden, J., 2008. A GIS-based assessment of coal-based hydrogen infrastructure deployment in the state of Ohio. Int. J. Hydrog. Energy 33, 5287–5303.

Kantar, Y.M., Usta, I., 2008. Analysis of wind speed distributions: wind distribution function derived from minimum cross entropy principles as better alternative to Weibull function. Energy Convers. Manag. 49, 962–973.

Melendez, M., 2006. Transitioning to a hydrogen future: learning from the alternative fuels experience. National Renewable Energy Laboratory, Golden, CO Technical Report No. NREL/TP-540-39423.

Ni, J., Johnson, N., Ogden, J.M., Yang, C., Johnson, J., 2005. Estimating Hydrogen Demand Distribution Using Geographic Information Systems (GIS). National Hydrogen Association, Washington, DC.

Ouammi, A., Dagdougui, H., Sacile, R., Mimet, A., 2010. Monthly and seasonal assessment of wind energy characteristics at four monitored locations in Liguria region (Italy). Renew. Sust. Energ. Rev. 14, 1959–1968.

Ouammi, A., Sacile, R., Mimet, A., 2010. Wind energy potential in Liguria region. Renew. Sust. Energ. Rev. 14, 289–300.

Ouammi, A., Zejli, D., Dagdougui, H., Benchrifa, R., 2012. Artificial neural network analysis of Moroccan solar potential. Renew. Sust. Energ. Rev. 16 (7), 4876–4889.

Rehman, S., 2004. Wind energy resource assessment for Yanbo, Saudi Arabia. Energy Convers. Manag. 45, 2019–2032.

Rosyid, O.A., 2006. System-analytic safety evaluation of hydrogen cycle for energetic utilization (Doctoral dissertation). Otto-von-Guericke-Universität Magdeburg, Germany.

Shata, A.S.A., Hanitsch, R., 2008. Electricity generation and wind potential assessment at Hurghada. Renew. Energy 33, 141–148.

Ucar, A., Balo, F., 2009. Evaluation of wind energy potential and electricity generation at six locations in Turkey. Appl. Energy 86, 1864–1872.

Zhiyong, L., Xiangmin, P., Jianxin, M., 2010. Harm effect distances evaluation of severe accidents for gaseous hydrogen refueling station. Int. J. Hydrog. Energy 35, 1515–1521.

Zhiyong, L.I., Xiangmin, P.A.N., Jianxin, M.A., 2011. Quantitative risk assessment on 2010 Expo hydrogen station. Int. J. Hydrog. Energy 36, 4079–4086.

Chapter 6

Network Planning of Hydrogen Supply Chain

1. APPROACHES FOR THE PLANNING AND DESIGN OF HYDROGEN INFRASTRUCTURE

The transition to a hydrogen economy has been studied in various locations and reported in many research studies. Each of these plans treats particular aspects related to the transition to hydrogen economy. This may vary according to national or regional plans/roadmaps, specific policies (Hajimiragha et al., 2009), or specific environmental targets. The published research studies that analyze the hydrogen infrastructure can be categorized into following three methodological approaches. In the first one, researchers focus on design of the hydrogen supply using the mathematical optimization methods, usually the architecture of these studies consists of presenting the general mathematical formalization, then an application of the model is performed for a national or regional case study. The second approach is related to studies that explained spatial models and frameworks for the design of hydrogen infrastructure. These approaches usually are related to country or region specific. The third approach that can be implemented in the design phase of HSC is the use of transition scenario and plans. These studies may aggregate future hydrogen scenario evaluation and cost estimation.

1.1 Optimization Methods

A literature review shows that the most common approach in designing and modeling a HSC is optimization methods. Various optimization techniques such as linear programming, dynamic programming, multiobjective programming, stochastic programming, and multiperiod were used by researchers to design the HSC in a most effective way. The aim of such method is to find out the optimal configuration that responds to some criteria (economic, safety, environmental). The input of those models considers a set of options for the production, storage, and transportation, while the output is to determine the type, numbers, location and capacity of the production, storage, transportation, etc. (Fig. 6.1).

Hydrogen Infrastructure for Energy Applications. https://doi.org/10.1016/B978-0-12-812036-1.00006-8
© 2018 Elsevier Inc. All rights reserved.

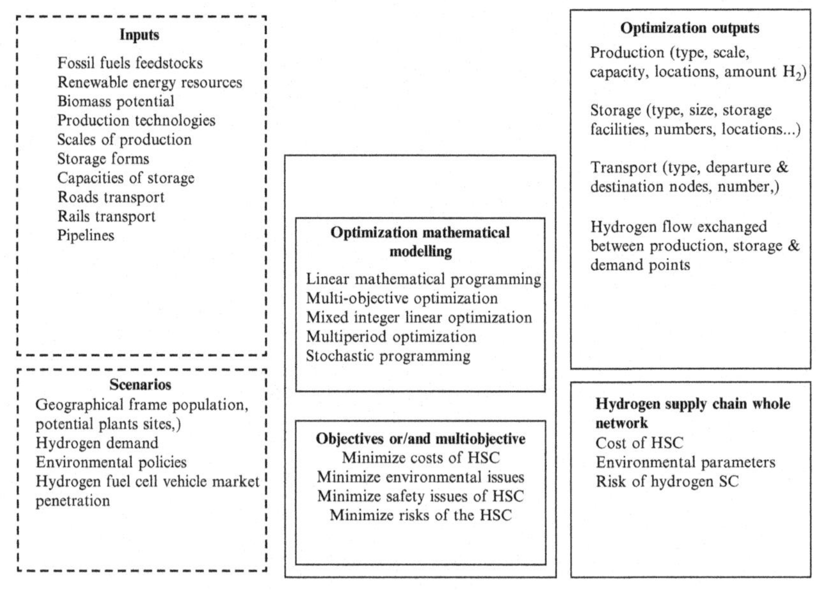

FIG. 6.1 Structure of the mathematical optimization of the HSC.

Almansoori and Shah (2009a) have proposed a high contribution to the design and planning of HSC network. Their models have been previously adopted by many authors all around the world. Their model consists of a multiperiod model for optimizing the operation of a future HSC. In their work, the authors have enhanced and gave more detail to their previous work which does not take into account many aspects, such as the primary energy source and the evolution of the network for the planning horizon. Ingason et al. (2008) presented a mixed integer linear programming (MILP) approach to locate the most economical site for hydrogen production technologies in Iceland. Authors' approach includes a feasibility study of exporting renewable energy in the form of hydrogen, from Iceland to Europe. The article discusses also how the total cost of hydrogen production can be minimized, based on costs of electricity production and transportation and hydrogen production. The results may provide ideas on which power plants to use, where to locate, and how to design the electrical transportation network in between. Brey et al. (2006) focused on the hydrogen economy and its development for Spain. The objective is to plan a gradual transition to a hydrogen economy solely based on the use of RES available within the territory having as target the satisfaction of hydrogen demand in a period of three years. Results have demonstrated that by the adoption of gradual transition, the GHG could drop by an amount of 4.54% by the target year. Also, from the security of supply perspective, authors have found that Spain region is self-sufficient for supplying hydrogen demand. Hugo et al. (2005) have developed a generic model for the optimal planning of future HSC for fuel cell vehicles. They used a mixed integer optimization techniques to find out optimal integrated investment strategies across a variety of supply chain decision-making stages. A concluding remark of the authors stated that to reach high GHG reduction target, the optimal supply chain design and investment strategy should start with on-site generation through small-scale reforming using natural gas from the grid. A method to determine the optimum hydrogen delivery network employing truck transportation has been developed in Kamarudin et al. (2009). Their method was based on the use of mixed integer linear programming to optimize a future hydrogen infrastructure in Malaysia. Authors have solved their model for two hydrogen demand calculation methods. Kim et al. (2008) have developed MILP model for the optimization of a hydrogen infrastructure under demand uncertainty. Authors proposed a stochastic method to take into account the effect of the uncertainty in the HSC. Comparing the two models with and without uncertainties, the main differences consist of the cost function, where various values can be obtained for each hydrogen demand value (source of uncertainty). Kim and Moon (2008a) have dealt with a multiobjective optimization approach considering cost and safety of HSC. The mathematical formulation of this model is an extension of the work by Kim et al. (2008)). The safety objective is here treated in terms of risk index, aiming essentially at the minimization of the population risk in the operation of the HSC. Konda et al. (2011) presented a multiperiod optimization that is based on a techno-economic analysis. Before introducing their model, authors have mentioned the hydrogen supply pathways through an overview of various parts of the hydrogen infrastructure. Then, the model formulated based on a MILP has been solved in the General Algebraic Modeling System Environment (GAMS). The approach was applied on a large-scale Dutch case study. The application of the model to the case study reveals that the transition toward a large-scale H_2-based transport is economically feasible, for any given demand scenario.

Guillén-Gosálbez et al. (2009) have studied bi-criteria mixed integer linear programming that targets to design a hydrogen network considering cost and environmental impact. In this work, authors focused on the analysis of the environmental impact from a life cycle analysis viewpoint. The authors have extended the model presented by Almansoori and Shah (2009a) in order to take into account the evolution of the network over time, especially considering time-variant demand. Sabio et al. (2010) have formulated a multiobjective problem to allow the control of variation of the economic performance of the hydrogen network. The author's approach is an extension of some previous published works. The design problem addressed in this work has an objective to determine the optimal configuration of a three-echelon HSC, namely, production, storage, and market, while minimizing the expected total discounted cost and the associated financial risk associated to market changes. Li et al. (2008) have presented and analyzed the case study of China under a multiperiod context to provide optimal integrated investment strategies across a variety of supply chain decision-making. Significant contributions appeared from the California state where hydrogen has started to be seen as an important alternative fuel and energy carrier. Parker et al. (2010) have constructed a model for finding the most efficient and economical configuration of the green energy pathway mainly based on agricultural residues. The approach is based on an integrated model that aims to evaluate the economic potential and the infrastructure requirements of the bio-hydrogen production from agricultural residues. The approach has been essentially applied to the northern California, where results have demonstrated that agricultural waste can be cost similar to the natural gas for the hydrogen production. Another approach has been applied this time to the southern California, where Lin et al. (2008) have investigated a model to determine the least cost hydrogen infrastructure design considering different technological alternatives. Lin et al. (2008) have developed a MILP model to optimize a hydrogen station siting in southern California by minimizing the fuel-travel-back time. Bersani et al. (2009) have investigated the planning of a network of service stations of a given company within a competitive framework. They proposed a decision support system that can be considered to determine the optimal placement of services stations within a hydrogen economy. Their method has been applied to a specific territory in northern Italy. Van den Heever and Grossmann (2003) have defined a mathematical model for a HSC. The study developed by Kim and Moon (2008b) has focused on a specific case related to the impact of introduction of hydrogen as fuel in the road transportation sector in Korea. The modeling has been done by the LEAP software which is an accounting and scenario-based modeling tool enabling the assessment of energy consumptions, required costs, and GHG emissions. Nicholas et al. (2004) have presented a model for locating hydrogen fuel stations assuming that the existing petrol infrastructure will be strongly related to the hydrogen infrastructure in the future. Melaina (2003) has discussed three approaches to estimate the number of hydrogen fueling stations that might be required to have convenient access to hydrogen fueling. These approaches are based upon the existing populations of gasoline stations, the metropolitan land areas, and the lengths of principal arterial roads. The author has demonstrated that the arterial roads approach appears to provide the most consistent analysis for both rural interstate and metropolitan area stations. Brey et al. (2006) have described a multiobjective optimization model, where the main aim is to satisfy around a certain percentage of energy demand for the transport in Spain by a certain time horizon. The main assumption of the study is the use of RES. Results permitted to determine for each region, what are the RES to be used to obtain hydrogen and hydrogen transport requirements between the regions.

1.2 Geographical Information System (GIS)-Based Approaches

It is important to underline that by contrast to the mathematical optimization approaches, the spatial or GIS approach cannot be considered as a general methodology for the finding of the optimal HSC configuration. In fact, the results of the approach are country/region-specific conditions, depending strongly on the local territorial condition, such as transportation network, population, available resources, local policies, and others. Fewer studies have focused on this framework; Stiller et al. (2009) have developed a GIS-based regional hydrogen demand scenarios and fueling station networks for the design of the pathways of hydrogen fuel in Norway. The authors' method considers growth of regional hydrogen coverage and the increase in the density of hydrogen users over time. Kuby et al. (2009) have presented a model that locates the hydrogen stations to fuel the maximum volume of vehicle flows. Their model includes a spatial decision support system using GIS coupled with a heuristic algorithm, which is used as a support tool to analyze different scenarios, evaluate trade-offs, and map the results. Johnson et al. (2008) have proposed a model for the optimization of a regional hydrogen infrastructure, thus combining two modules, namely, the special data in GIS and a techno-economic model of hydrogen infrastructure. Their results have demonstrated that by the aggregation of infrastructure at the regional scale yields lower levelized costs of hydrogen than at the city level at a given market penetration level. Also, they concluded that the centralized production with pipeline distribution is the favored pathway even at low market penetration. From the US National Renewable Energy Laboratory (NREL), Melendez and Milbrandt (2008) have developed a GIS-based study for the identification of the minimum hydrogen infrastructure to gain consumer buy-in for purchasing hydrogen vehicles in the United States. Authors have also presented a GIS method for siting hydrogen stations, thus based on the demand characteristics of select urban areas. The above cited studies mainly deal with papers that have implemented solely the geographical information system to design the hydrogen infrastructure. However, additional enhancements could be projected by coupling the GIS-based module to additional mathematical models which lead to an integrated approach. This coupling could favor the exploitation of two different decision support systems. For instance, Strachan et al. (2009) have described an integrated approach linking spatial GIS modeling of hydrogen supply, demands, and infrastructures, anchored within an economy-wide energy systems model (MARKAL). The study was specifically applied to the United Kingdom (UK). Ball et al. (2007) have proposed a plan for the integration of the hydrogen economy into the German energy system. The objective of the modeling approach is to optimize—for an exogenously given, regionally distributed hydrogen demand—the buildup of a hydrogen infrastructure over space and time, and to assess the corresponding economic and environmental effects. A model was developed as a novel tool to assess the introduction of hydrogen as vehicle fuel by means of an energy system analysis.

1.3 Evaluation Plans Toward the Transition to Hydrogen Infrastructure

While some authors have developed mathematical and GIS-based approaches, others have presented transition models to the future HSC. The objective here is not to model the hydrogen infrastructure but to understand the behavior of the chain in certain areas assuming specific scenarios. Usually, these kinds of studies are accompanied with the cost estimation of the hydrogen pathways. These transition models are implemented on a country or region basis, aggregating simultaneously territorial information and specific data (such local policies and regulations). Lee et al. (2009) have studied the environmental aspects of hydrogen pathways in Korea. The objective of this paper is to evaluate the environmental aspects of hydrogen pathways according to hydrogen production methods, production capacities, and distribution options. The methodology applied to reach the target is the life cycle assessment (LCA). Results of the LCA applied to the Korean case study show that wind is superior regarding its potential in the reduction of global warming. In fact, authors have demonstrated that the substitution of gasoline with wind energy can reduce the global warming and fossil fuel consumption by 99%. Farrell et al. (2003) have reviewed different strategies for the introduction of hydrogen as a transportation fuel. Authors claimed that the cost of introducing hydrogen can be reduced through the selection of a mode that uses a small number of relatively large vehicles that operate along a limited number of point-to-point routes or within a small geographic area. From an environmental point of view, the authors suggested that the environmental benefits of hydrogen uses as a fuel can be reached through the introduction in the modes that have little or no pollution regulations that is applied to them. At the European scale, an estimation based on hydrogen penetration scenarios in Europe has been carried out by Tzimas et al. (2007). The authors have evaluated the evolution in the size and the cost of the hydrogen delivery infrastructure in Europe. The estimation study has showed that between 1 and 4 million km of pipelines distribution may be needed. While, a cumulative capital between 700 and 2200 thousand million euros is necessary to build the infrastructure by 2050. In a similar study, Wietschel et al. (2006) have presented a European hydrogen infrastructure in Europe by 2030. The study was based on the hydrogen penetration level in Europe, where two scenarios have been considered. The decisions related to the selection of HSC were mainly based on both the energy chain calculation costs, emissions, and the expert judgments among others. It can be shown that under economic and CO_2-reduction objectives, the steam reforming of gas is the primary most promising hydrogen production options in this first phase for developing a hydrogen infrastructure. Ogden (1999) has examined the techno-economic feasibility of developing an infrastructure for the hydrogen with zero emissions vehicles. The modeling has been applied to the southern California. Different possibilities for producing and delivering gaseous hydrogen transportation have been analyzed. Shane and Samuelsen (2009) have discussed a novel tool entitled Preferred Combination Assessment (PCA) that enables the analysis of the impacts of an integrated HSC. This latter tool will allow the search for a HSC that respects the criteria pollutant emissions, GHG emissions, and energy utilization. The inputs of PCA are among others the total hydrogen demand in a region, number of hydrogen refueling stations, production facilities, and distances over which it must be delivered, whereas the outputs are the criteria pollutant emissions, GHG emissions, energy consumption, and water consumption. Hake et al. (2006) have reviewed the prospects of hydrogen in the energy system considering various scenarios. Authors have first reviewed the hydrogen experiences available in Germany, such those related to the infrastructure for the transport of hydrogen (pipeline, railways, and road trailers). Then, they have considered the prospects of three applications of hydrogen, namely, stationary, mobile, and portable power supply. They claimed that the role of hydrogen will be small in the coming decades mainly due to the high investment costs. One solution that might be advantageous for the hydrogen introduction is through its development under constrained conditions. Smit et al. (2007) have presented an excel simulation for the study of the transition to hydrogen energy. In particular, the main objective is to quantify the Dutch hydrogen transition evaluating, for example, the prospects of hydrogen on-site production and its role for the developing of the hydrogen demand. Authors have found that the use of locally produced hydrogen from natural gas in stationary and mobile applications can yield an economic advantage when compared to the conventional system and can hence generate hydrogen demand. Contaldi et al. (2008) have analyzed the hydrogen market in Italy that needs to be established in order to meet climate change, environmental, and energy security issues. The study was done following an Italy-Markal model. Different hydrogen technologies were considered, while the transportation was the only end user. Kruger et al. (2003) have explored the potential in New Zealand for the use of hydrogen as a transportation fuel. Their study was based on the use of some historical data on vehicle transport, population, and electric energy to make the estimations regarding the requirements of hydrogen fuel for a certain projected horizon.

1.4 Conclusion

A classification of literature related to the HSC has been done, and which has distinguished three main classes of paper, namely, mathematical optimization methods, GIS-based approaches, and assessment plans for the planning of HSC. Based on the review, it is observed that many authors have focused on design/planning of HSC using mathematical optimization methods. These methods are the most effective ones to best address the question of future hydrogen infrastructure design. Main objectives to be minimized within the optimization of HSC are related to cost and environment. Fewer studies have addressed the optimization of the HSC from the hydrogen risks viewpoint. Considering the state-of-the-art HSC, it is also identified that more researches are needed in addressing HSC that operate on clean feedstocks, such those based on RES. For this main reason, the aim within this thesis is to investigate a hydrogen supply chain based on the use of renewable energy sources. The design of such a Green Hydrogen Supply Chain will be based on the use of the optimization methods which have the potential to design a general network of hydrogen infrastructure that can be applied to a variety of case studies. In the optimization of such infrastructure, we will focus also on the risk side aiming to select an infrastructure that minimizes the risks on the population and environment. In this frame, the focus will be dedicated to the evaluation of the technical feasibility and the performance of renewable HSC. From the literature review, a GIS-based methodology will be applied based on a specific case study. The GIS methodology—in contrast to which is published—will be considered to analyze the clean feedstock for hydrogen production and further the tool will support in designing a decision support system to select better sites for hydrogen production. One additional component that will be considered in this thesis is the risk criteria which were not considered enough in the design of a future hydrogen infrastructure, so a design implementing the risk criteria will be studied and considered.

2. MODELING AND CONTROL OF HYDROGEN IN A NETWORK OF GREEN HYDROGEN REFUELING STATIONS

2.1 Introduction

The global awareness concerning greenhouse gas (GHG) emissions, air pollution, fossil fuel depletion, and other energy security issues (Li et al., 2009; Demirbas, 2008) have led many governments and researchers around the world to develop secure and environmental friendly fuel. The current fossil fuel systems must be switched gradually to clean, affordable, and reliable energy systems, thus to reach the global drivers for a sustainable vision of our future energy market. Among many alternative energy sources, hydrogen can be considered as an attractive solution to succeed the current carbon-based energy system. The main benefits of hydrogen are even substantially considered by the fact that hydrogen can be manufactured from a number of primary energy sources, such as natural gas, nuclear, coal, biomass, wind, and solar energy. Such diversity in production, obviously contributes significantly in diversifying the energy supply system and in ensuring the security of fuel supply. For transport applications, there will be an increasing requirement to use clean and low or zero emission fuels such as hydrogen (Clarke et al., 2009). In addition, hydrogen is the most abundant element on the earth; it is clean and has the highest specific energy content of all conventional fuels (Campen et al., 2008). Hydrogen can contribute to a diversification of automotive fuel sources and supplies and can offer long-term solution being solely produced from renewable energies. The development of a hydrogen infrastructure for producing and delivering hydrogen appears as a key factor to achieve the hydrogen economy transition and its development. In fact, the modeling of hydrogen infrastructure is still a complex task; the main complexities rise from the significant uncertainties in demand; supply; economic and environmental impacts; and in the diversity of technologies available for production, storage, and transportation. The key question is from which sources hydrogen can be produced in a sustainable manner (Ingason et al., 2008). The extent to which the hydrogen benefits will occur has a great dependency on the technologies involved. Many authors have agreed that renewable energy sources (RES), such as wind and solar are central for better transition to a long-term hydrogen economy. In order to reach that goal, it is advantageous to use renewable energy for hydrogen generation. In fact, the resources for the operation of renewable energy systems are inexhaustible and practically free making. In addition, sustainable hydrogen production from electrolysis yields several advantages from a system point of view (Jørgensen and Ropenus, 2008). For instance, wind-powered water electrolysis ranks high in terms of technical and economic feasibility, having a great potential to become the first competitive technology to produce large amounts of renewable hydrogen in the future (Sherif et al., 2005; Bokris and Veziroglu, 2007; Greiner et al., 2008; Segura et al., 2007; Prince-Richard et al., 2005).

From the end users' perspective, the use of hydrogen in fuel cell applications offers a number of advantages over existing fuels and other emerging competitors, especially in the transportation sector (Hugo et al., 2005). The fuel cell vehicles can be a long-term solution to the persistent environmental problems associated with transportation. The fuel cell vehicles would be less complex, have better fuel economy, lower GHG emissions, greater oil import reductions, and would lead to a sustainable transportation system once renewable energy is used to produce hydrogen (Thomas et al., 1998). According to Doll and Wietschel (2008), the introduction of hydrogen coupled with the fuel cell vehicles could reduce significantly the emissions of CO_2, NOx, and SOx.

The transition to a sustainable hydrogen economy faces paramount economic and technological barriers that must be overcome in order to ensure a successful transition. It is essential to study and analyze the interactions between different hydrogen infrastructure components in advance in order to set and build a variety of options for the incorporation of this new economy. This will facilitate the management of hydrogen supply chain and help decision makers to define adequate roadmaps for the hydrogen development.

Many authors have detailed approaches and models for the development of the future hydrogen infrastructures. The approaches range from the examination of the supply chain as a whole (Kim et al., 2008; Almansoori and Shah, 2009b; Almansoori and Shah, 2006) to the focus on a node of the infrastructure such as production, storage, or transportation (Sherif et al., 2005; Parker et al., 2009; Joffe et al., 2004; Dagdougui et al., 2010; Martin and Grasman, 2009). Kuby et al. (2009) have developed a model able to locate the hydrogen stations that refuel maximum volume of vehicle fuel, this latter is measured both using the number of trips and vehicle miles travelled. Bersani et al. (2009) have investigated the planning of a network of service stations of a given company within a competitive framework. They proposed a decision support system that can be considered to determine the optimal placement of service stations within a hydrogen economy. Nicholas et al. (2004) have provided an analytical framework for locating hydrogen fuel stations assuming that the existing petrol infrastructure is strongly related to the needed hydrogen infrastructure of the future. In another study, Parker et al. (2009) have assessed the economic and infrastructure requirements of the production of hydrogen from agricultural wastes; they concluded that the delivery price of bio-hydrogen is similar to the hydrogen produced from the natural gas. Joffe et al. (2004) have developed a technical modeling of a hydrogen infrastructure. They investigated the operation of the system so as to provide initial facility for refueling hydrogen fuel cell buses in London city. Greiner et al. (2008) have presented a simulation study of combined wind-H_2 plant on a small Norwegian island. They included chronological simulations and economic calculations enabling the optimization of the components size. Their simulations include a grid-connected system and an isolated system with backup power generator. Dagdougui et al. (2010) have introduced a dynamic decision model for the real-time control of hybrid renewable energy production systems, which can be particularly suitable for autonomous systems.

General interest in a wind-hydrogen system has increased partly because the price of wind power has become competitive with traditional power generating sources in certain areas (Martin and Grasman, 2009). Wind-powered water electrolysis ranks high in terms of technical and economic feasibility, having a great potential to become the first competitive technology to produce large amounts of renewable hydrogen in the future (Sherif et al., 2005). Worldwide installations of wind turbine power have reached a value of 194.5 GW (Syste‘mes solaires le journal de l’e´olien n_ 8, 2011). Studies carried out by Honnery and Moriarty (2009) have evaluated the global potential of a coupled wind-hydrogen system, thus in order to estimate the future hydrogen production.

A challenging task that is worth to be deeply studied regards the feasibility to feed hydrogen demand points to an uncertain renewable supply, such as the case of hydrogen production from intermittent RES. The key question that needs to be addressed is the ability of the renewable energy system to meet the hydrogen fuel requirements (in amount and time).

In this section, an attempt has been made to plan an innovative design of a hydrogen infrastructure. It consists of a network of GHRSs and several production nodes. The proposed model is formulated as a mathematical programming, where the main decisions are the selection of GHRSs that will be powered by each point of production based on distance and population density criteria, as well as the energy and hydrogen flows exchanged among the system components from the production nodes to the demand points. The approaches and methodologies developed can be taken as a support to decision makers, stakeholders, and local authorities in the implementation of future hydrogen infrastructures.

2.2 The Methodological Approach

2.2.1 Model Structure and Components

Wind/solar energy production systems are designated to supply electric and hydrogen needs to a network of GHRS. From the demand viewpoint, the system to be modeled consists of a network of GHRS, each one having a local hydrogen storage tank, and needs to procure hydrogen to costumers (in terms of fuel cell vehicles). From the supply viewpoint, production systems for hydrogen consist of many mixed energy production plants. Each one includes a large-scale hybrid wind/solar system, electrolyzer unit, fuel cell, and a main hydrogen storage tank. Moreover, each production plant is connected to the electrical network, giving the system the possibility to sell excess of energy generated by the production nodes. Fig. 6.2 displays the proposed hydrogen infrastructure that forms the basis of the hydrogen global system.

2.2.2 Planning Horizon: Plants and Technologies

The implementation of planning phase is important for the formulation of the optimization model, because it will provide the knowledge about the possibility

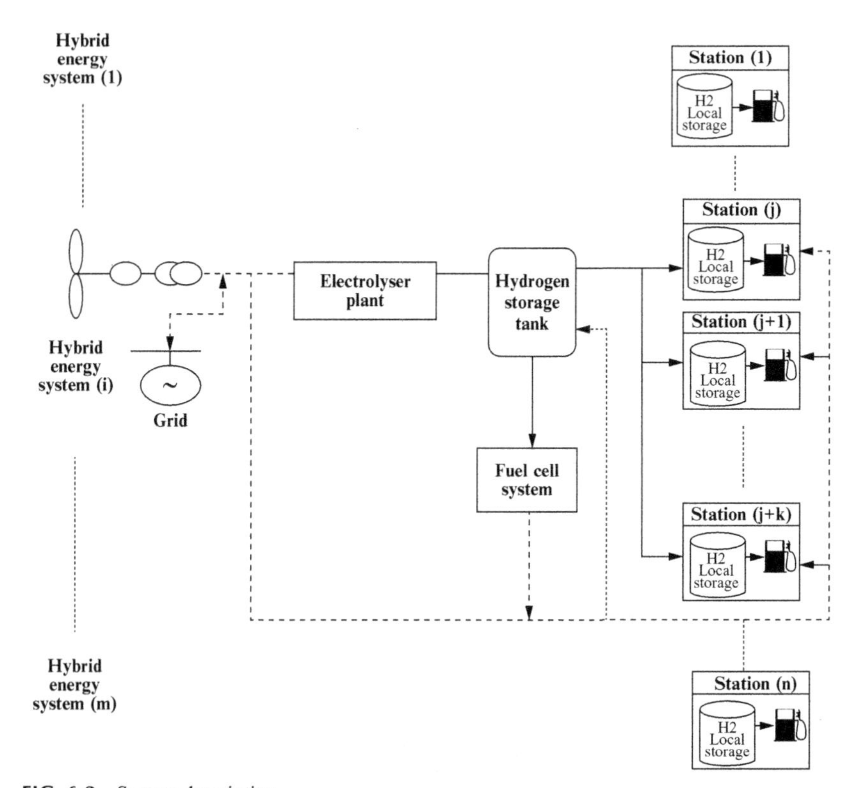

FIG. 6.2 System description.

of making use of the renewable resources for hydrogen production. Furthermore, due to the absence of large hydrogen market, the development of the hydrogen infrastructure will initiate by the introduction of small hydrogen decentralized systems. Hence, the operation of these systems as a whole will depend, in addition to the available renewable resources, to the design of the system that will generate and fulfill specific requirements. In fact, the system must be adapted accurately in order to fit the mass of hydrogen for the GHRS. Nevertheless, economically speaking, the configuration of the best design will strongly depend on the considered scenario. In other words, on the renewable resources condition as well as the cost of the different components and the final selling prices of the energy generated by each type of renewable source (López et al., 2009). The main idea of using mixed renewable energy systems for hydrogen production is based on generating electrical energy by the wind turbines and the photovoltaic modules, and then using that energy for hydrogen generation using electrolyzer systems. Owing to the low energy density by volume of the hydrogen, the effective use of the logistics chain to transport hydrogen to the refueling stations requires the reduction of the hydrogen volume, thus by means of liquefaction or compression. In this paper, liquefaction or compression of the hydrogen substance will not be considered, neither the distributing chain of hydrogen to the refueling stations.

2.2.2.1 Mixed Renewable Energy System

The selection of the suitable renewable energy technologies has been made on the basis of several considerations: meteorological conditions, renewable energy potential, load factor, and amount of hydrogen required by the refueling stations. The problem needs much more attention, especially because of the use of the intermittent RES, which is mainly the case of solar and wind. Due to these issues, the mixed renewable energy system must be well designed in order to better exploit the potential of RES, hence fulfill specific requirements.

2.2.2.2 Electrolyzer Unit

The hydrogen generation process becomes more beneficial if used in conjunction with electricity generated by the RE. Different kinds of water electrolyzers exist, with various levels of technological advancements. Water electrolyzers can be divided into two categories, alkaline and proton exchange membrane (PEM) electrolyzers. PEM electrolysis is a viable alternative for generating hydrogen from RES. Despite its higher cost compared, the PEM benefits of high efficiency factor, higher life cycle (approximately ten years) (Benchrifa et al., 2007), adaptability with renewable energy systems (Kruse et al., 2002). In addition, a PEM electrolyzer can deliver hydrogen at high pressure, which will in turn be attractive for the application where hydrogen needs to be stored (Barbir, 2005). In this study, a PEM electrolyzer will be adopted, this choice is justified by two main reasons, namely, the exploitation of wind/solar potentials and the need of high pressure hydrogen to refueling stations.

2.2.2.3 Fuel Cell System

Among all kinds of fuel cell available, fuel cell proton exchange membrane PEM will be the most suitable for the system configuration; this choice is dictated by the PEM electrolysis. The operation of PEM fuel cell needs only hydrogen, oxygen from the air, and water to operate and do not require corrosive fluids like some fuel cells. The driven goal of using a fuel cell system at the centralized plant is to ensure another option for the provision of electricity (in case of low power generation from mixed renewable energy system).

2.2.2.4 Hydrogen Storage System

A hydrogen storage system is available in each hydrogen production plant. Since the plant is supposed to provide hydrogen for a network of hydrogen refueling station, there is a need to assume the availability of a large-scale hydrogen storage system. The storage system assumes the aggregation of both a compressor and storage system. The hydrogen storage device has to be able to meet the requirements of the hydrogen demand at the specific hydrogen refueling station in each type (taking into consideration the delivery schedule). Because of lack of data in particular those related to large-scale hydrogen storage. The hydrogen storage systems have been sized by analyzing the real amount of gasoline delivered by several service stations which is transformed in equivalent amount of hydrogen.

2.2.2.5 Hydrogen Demand

Due to the absence to date of a widespread hydrogen market, the computation of the hydrogen demand cannot be estimated by accuracy. The operation of these markets will depend initially on the hydrogen demand in the available infrastructure, in particular, the number of hydrogen fuel cell vehicles and hydrogen refueling infrastructure. In addition, due to many uncertainties, the planning of scenario can be considered as the only systematic tool that helps designing the hydrogen supply chain. The configuration of the optimal design will strongly depend on the scenario considered. In this study, the estimation of hydrogen demand of the hydrogen refueling stations is performed based on the current supply of the petrol products to the conventional petrol stations. The information related to the petrol stations may be helpful in determining the capacity and consumption of the future hydrogen refueling stations. Then, according to these estimations, multiple scenarios are explored including varying hydrogen fuel cell penetrations.

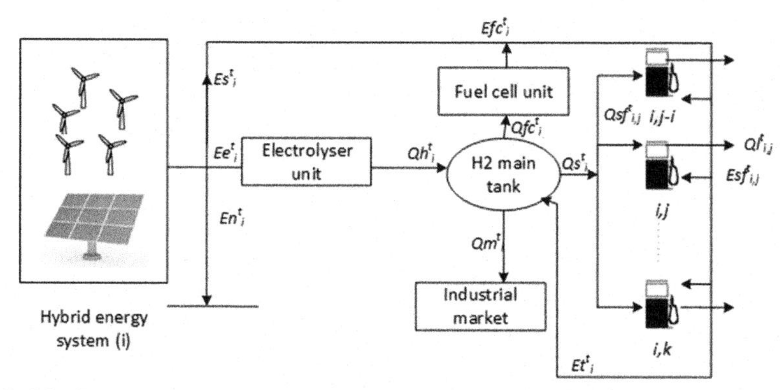

FIG. 6.3 Energy and hydrogen flows exchanged among system components.

2.2.3 Modeling the Mixed Renewable Energy System

As presented previously, the electrolysis hydrogen production plant is mainly driven by power generated from the renewable energy system. The overall hydrogen production plant is linked to the network of GHRS. It is composed of the following subsystems: photovoltaic modules, wind turbines, electrolyzer unit, fuel cell unit, main hydrogen storage system. Among these subsystems and from/to the external electrical network, the power and hydrogen flows can be exchanged as shown in Fig. 6.3. The energy produced by the renewable energy system can be sent to the electrolyzer to produce hydrogen that will be stored in main tank and/or used to feed electrical energy of GHRS. The quantity of hydrogen can be used for different purposes: to satisfy the hydrogen fuel demand by the refueling stations and/or used by the fuel cell to ensure the electricity demand of the refueling stations in the case of a deficit of energy production, and/or delivered to the industrial markets.

2.2.3.1 The Wind Turbine Subsystem Model

The energy produced by a wind turbine in a point of production i is given by:

$$E_{w,i}^{t} = \frac{T}{1000} \frac{\rho A}{2} \int_{vi}^{vp} c_p \eta_{gb} \eta_g f\left(v_i^t\right) v_i^{t3} dv, \quad t = 1, \ldots, T \tag{6.1}$$

where $f(v_i^t)$ is the probability of occurrence of wind speed v_i^t (m/s) in a point of production i, A (m^2) is the blade sweep area, ρ is the air density (kg/m^3), c_p (–) is the power coefficient, η_{gb} (–) is the gearbox performance, and η_g (–) is the generator performance.

Where v_i^t corresponds to the wind speed at the wind turbine hub height. It is assumed that the wind speed can be predicted by some reliable meteorological models. In general, the wind speed measurements are given at a height different than the hub height of the wind turbine. So, the following equation is used to evaluate the wind speed at the desired height:

$$v_i^t = v_{data,i}^t \frac{\ln(H_{hub}/z_0)}{\ln(H_{data,i}/z_0)}, \quad t=1,\ldots,T \tag{6.2}$$

where $H_{data,i}$ (m) is the height of the measurement, H_{hub} (m) is the hub height, and z_0 is the surface roughness length and $v_{data,i}^t$ is the wind speed at the height of the measurements.

2.2.3.2 The PV Module Subsystem Model

Hybrid energy systems are often taken into account as a viable approach to face the RES intermittent character. The use of different RES (such as wind and solar) can enhance the effectiveness to face the load energy demands. The electrical energy generated from a photovoltaic module can be calculated using the following formula:

$$E_{pv,i}^t = S_{pv}\eta_{PV}p_f\eta_{pc}G_i^t, \quad t=1,\ldots,T \tag{6.3}$$

where S_{pv} (m^2) is the solar cell array area, η_{PV} (–) is the module reference efficiency, p_f (–) is the packing factor, η_{pc} (–) is the power conditioning efficiency, and G_i^t (kWh/m^2) is the forecasted hourly irradiation that is predicted by some reliable meteorological model.

2.3 Optimization Problem

The decision variables of the optimization problem are as follows:

Ee_i^t, En_i^t, and Es_i^t (kWh) are the electrical energy components in time period $[t, t+1)$, respectively, given by the mixed renewable energy system i to electrolyzer unit, electrical network, and to feed the electrical demand of the GHRS.

Efc_i^t, Et_i^t, and $Esf_{i,j}^t$ (kWh) are the electrical energy components in time period $[t, t+1)$, respectively, given by the fuel cell, consumed by the main hydrogen storage tank in point of production i and consumed by the GHRS j and its local hydrogen tank.

Qh_i^t, $Qsf_{i,j}^t$, $Ql_{i,j}^t$, Qm_i^t (kg) represent in time interval $[t, t+1)$, respectively, the amount of hydrogen produced by the electrolyzer unit at production node i, the amount of hydrogen sent from the ith production node to the jth GHRS, the amount of hydrogen delivered by the jth GHRS, and the amount of hydrogen sent to the industrial market from the ith production node.

Where $\rho_{i,j}$ is a binary variable, equals to 1 if the ith production point will be connected to the jth GHRS, 0 otherwise.μ_j is a binary variable, equals to 1 if the jth refueling station is selected, 0 otherwise.

The state variables of the optimization problem are as follows:

M_i^t, $Msf_{i,j}^t$ are, respectively, the amount of hydrogen stored in time period $[t, t+1)$ in the main tank of ith production point and the amount of hydrogen stored in the local tank of the jth GHRS at instant t.

The following parameters are used in the model

- N_w: number of wind turbines;
- N_{pv}: number photovoltaic modules;
- HHV_{H_2}: hydrogen higher heating value (kWh/kg)
- LHV_{H_2}: hydrogen lower heating value (kWh/kg)
- LHV_{ff}: fossil fuel lower heating value (MJ/kg)
- η_1, η_2: parameters of the electrolyzer plant: the first one represents the efficiency of the electrolysis system, while the second one is an additional efficiency coefficient included to take into account the (energy) losses in the electrolyzer.
- η_{ff}: efficiency of the fossil fueled engine
- η_{H_2}: efficiency of the H_2 engine/fuel cell (–)
- η_{FC}: fuel cell efficiency (–)
- $Qff_{i,j}^t$: demand of fossil fuel in time interval $[t, t+1)$ at the jth refueling station (kg)
- $\widetilde{Qm}_i^t$: hydrogen market demand in time interval $[t, t+1)$ requested from the ith production point (kg)
- $\widetilde{Ql}_{i,j}^t$: hydrogen demand in time interval $[t, t+1)$ of the jth GHRS supplied by ith production point (kg)
- $\widetilde{Ee}_{i,j}^t$: electrical demand in time interval $[t, t+1)$ of the jth GHRS connected to the ith production point (kWh)
- $D_{i,j}$: distance from the ith production point to the jth GHRS (m)
- N_j: population density around a circle of unit diameter of distance of jth GHRS (inhabitant/km^2)
- δ_i: binary variable that is implemented to take into account the exchange in power between the hydrogen production plant and the electric network. It is equal to 1 when the market penetration of hydrogen is less enough to allow power exchange with the electrical network.

2.3.1 Objective Function

The objective function to be minimized is the sum of seven terms: two first terms reflect the satisfaction of different hydrogen demands, namely, the hydrogen fuel demands of the GHRS and the industrial market demand. The third term is related to ensure the electrical demands of GHRS, the fourth term is related to the energy sold to the electrical network to be maximized/minimized according to the market penetration, it means, for low hydrogen demand the energy sold to the electrical network will be maximized, otherwise it will be minimized, this assuming that the renewable energy systems are well sized. The fifth term to be maximized is related to the amount of hydrogen stored in the main hydrogen tank. The two last terms are added to select the suitable GHRS to be powered by the appropriate point of productions.

$$
\begin{aligned}
Z = & \sum_{t=1}^{T-1}\sum_{j=1}^{J}\left(Ql_{i,j}^{t} - \widetilde{Ql}_{i,j}^{t}\right)^{2} + \sum_{t=1}^{T-1}\sum_{i=1}^{I}\alpha\left(Qm_{i}^{t} - \widetilde{Qm}_{i}^{t}\right)^{2} \\
& + \gamma\sum_{t=1}^{T-1}\sum_{j=1}^{J}\sum_{i=1}^{I}\left(E_{i,j} + \varphi Msf_{i,j}^{t} - \widetilde{Ee}_{i,j}^{t}\right)^{2} + \zeta\sum_{t=1}^{T-1}\sum_{i=1}^{I}\delta_{i}En_{i}^{t} - \lambda\sum_{i=1}^{I}M_{i}^{T-1} \\
& + \beta\sum_{j=1}^{J}\sum_{i=1}^{I}\rho_{i,j}D_{i,j}^{2} + \chi\sum_{j=1}^{J}\mu_{j}N_{j}, \quad t=1,\ldots,T-1;\ i=1,\ldots,I;\ j=1,\ldots,J
\end{aligned} \tag{6.5}
$$

α, γ, ζ, λ, β, and χ are weight factors.

2.3.2 Constraints

2.3.2.1 Flow Conservation

The overall energy produced $E_{tot,i}^{t}$ (kWh) by each mixed renewable energy system in time interval $[t, t+1)$ is given by:

$$E_{tot,i}^{t} = N_{w} \cdot E_{w,i}^{t} + N_{pv} \cdot E_{pv,i}^{t}, \quad t=1,\ldots,T-1;\ i=1,\ldots,I \tag{6.6}$$

The total energy $E_{tot,i}^{t}$ (kWh) can be used for three different purposes in the same time interval: direct for the hydrogen production (Ee_{i}^{t}), satisfy the electricity demand of the refueling stations (Es_{i}^{t}), and a part sent directly to the electrical network (En_{i}^{t}). Thus

$$E_{tot,i}^{t} = Ee_{i}^{t} + En_{i}^{t} + Es_{i}^{t}, \quad t=1,\ldots,T-1;\ i=1,\ldots,I \tag{6.7}$$

$$Es_{i}^{t} + Efc_{i}^{t} - Ees_{i}^{t} - Et_{i}^{t} = 0, \quad t=1,\ldots,T-1;\ i=1,\ldots,I \tag{6.8}$$

The electrical energy Ees_{i}^{t} (kWh) consumed in time interval $[t, t+1)$ by kth GHRSs will be equal to the sum of electrical energy consumed by each one, and it is given by:

$$Ees_{i}^{t} = \sum_{j=1}^{k<J} Esf_{i,j}^{t}, \quad t=1,\ldots,T-1;\ i=1,\ldots,I;\ j=1,\ldots,J \tag{6.9}$$

$$Esf_{i,j}^{t} = E_{i,j} + \varphi Msf_{i,j}^{t}, \quad t=1,\ldots,T-1;\ i=1,\ldots,I;\ j=1,\ldots,J \tag{6.10}$$

where $Esf_{i,j}^{t}$ (kWh) is the electrical energy consumed in time interval by the jth GHRS that is composed by a constant term $E_{i,j}$ and a variable one $\varphi Msf_{i,j}^{t}$, with φ the unitary demand of electric energy for a unit of hydrogen stocked reservoir, and $Msf_{i,j}^{t}$ the amounts of hydrogen available in the jth local tank of the GHRS in time interval $[t, t+1)$.

The electrical energy Et_i^t (kWh) consumed in time interval by the main hydrogen storage reservoir in the point of production i is given by:

$$Et_i^t = \varphi M_i^t, \quad t=1,\ldots,T-1;\ i=1,\ldots,I \tag{6.11}$$

where M_i^t is the amount of hydrogen available in the main storage reservoir in time interval $[t, t+1)$.

The electrical energy Efc_i^t (kWh) delivered by the fuel cell in time interval $[t, t+1)$ is calculated by:

$$Efc_i^t = LHV_{H_2} \cdot Qfc_i^t \cdot \eta_{fc}, \quad t=1,\ldots,T-1;\ i=1,\ldots,I \tag{6.12}$$

The amount of hydrogen Qh_i^t (kg) delivered in time interval $[t, t+1)$ by the electrolyzer plant is equal to:

$$Qh_i^t = \frac{\eta_1 \eta_2 Ee_i^t}{HHV_{H_2}}, \quad t=1,\ldots,T-1;\ i=1,\ldots,I \tag{6.13}$$

The amount of hydrogen Qs_i^t consumed in time interval $[t, t+1)$ by the kth GHRS is equal to the sum of hydrogen that is consumed by each one, and it is given by:

$$Qs_i^t = \sum_{j=1}^{k<J} Qsf_{i,j}^t, \quad t=1,\ldots,T-1;\ i=1,\ldots,I;\ j=1,\ldots,J \tag{6.14}$$

The hydrogen demand $\widetilde{Ql}_{i,j}^t$ (kg) of the jth GHRS in time interval $[t, t+1)$ is given by:

$$\widetilde{Ql}_{i,j}^t = \frac{Qff_{i,j}^t \cdot LHV_{ff} \cdot \eta_{ff}}{LHV_{H_2} \cdot \eta_{H_2}}, \quad t=1,\ldots,T-1;\ i=1,\ldots,I;\ j=1,\ldots,J \tag{6.15}$$

2.3.2.2 Storage System State Equations

The state equations of the main hydrogen storage tank of each point of production and the local hydrogen storage tank of each GHRS in time interval $[t, t+1)$ are given by:

$$M_i^{t+1} = M_i^t + Qh_i^t - Qs_i^t - Qm_i^t - Qfc_i^t, \quad t=1,\ldots,T-1;\ i=1,\ldots,I \tag{6.16}$$

$$Msf_{i,j}^{t+1} = Msf_{i,j}^t + Qsf_{i,j}^t - Ql_{i,j}^t, \quad t=1,\ldots,T-1;\ j=1,\ldots,J \tag{6.17}$$

$$M_i^1 = M_{i,1}, \quad i=1,\ldots,I \tag{6.18}$$

$$Msf_{i,j}^1 = Msf_{i,j,1}, \quad i=1,\ldots,I;\ j=1,\ldots,J \tag{6.19}$$

where $M_{i,1}$, $Msf_{i,j,1}$ (kg) are, respectively, the storage system level at the initial time for each main hydrogen storage tank and each local one.

2.3.2.3 Other Constraints

Selection GHRS Constraint (19) is introduced in order to impose that if the link between the ith production point and the jth GHRS is established. So, the jth GHRS must be selected following the safety viewpoint.

$$\text{If } \exists\ \rho_{i,j}=1 \rightarrow \gamma_j=1$$

$$\rho_{i,j}=\begin{cases}1 & \text{if the link}\,(i,j)\,\text{is established}\\ 0 & \text{otherwise}\end{cases} \tag{6.20}$$

Eq. (6.20) can be also written as:

$$\rho_{i,j}-R\gamma_j\leq 0 \tag{6.21}$$

where R is a big number

Hydrogen Storage Systems The main hydrogen storage tanks are limited in upper and lower bands:

$$M_{i,\min}\leq M_i^t\leq M_{i,\max},\quad t=1,\ldots,T-1;\ i=1,\ldots,I \tag{6.22}$$

The local hydrogen storage tanks are limited in upper and lower bands:

$$Msf_{i,j,\min}\leq Msf_{i,j}^t\leq Msf_{i,j,\max},\quad t=1,\ldots,T-1;\ j=1,\ldots,J \tag{6.23}$$

Activation of Transport The hydrogen flows sent from a supply point i to the jth GHRS are limited in upper and lower bands:

$$\rho_{i,j}Qsf_{j,\min}\leq Qsf_{i,j}^t\leq \rho_{i,j}Qsf_{j,\max},\quad t=1,\ldots,T-1;\ i=1,\ldots,I;\ j=1,\ldots,J \tag{6.24}$$

Electrolyzer The energy flows sent to the electrolyzer unit at each production point i are limited in upper and lower bands:

$$\delta_e^t Ee_{i,\min}\leq Ee_i^t\leq \delta_e^t Ee_{i,\max},\quad t=1,\ldots,T-1;\ i=1,\ldots,I \tag{6.25}$$

$$\delta_e^t=\begin{cases}0 & \text{if}\ \ Ee_i^t<Ee_{i,\min}\\ 1 & \text{if}\ \ Ee_i^t\geq Ee_{i,\min}\end{cases} \tag{6.26}$$

Fuel Cell The hydrogen flows sent from the fuel cell are limited in upper and lower bands:

$$\delta_{fc}^t Qfc_{i,\min}\leq Qfc_i^t\leq \delta_{fc}^t Qfc_{i,\max},\quad t=1,\ldots,T-1;\ i=1,\ldots,I \tag{6.27}$$

$$\delta_e^t=\begin{cases}0 & \text{if}\ \ Qfc_i^t<Qfc_{i,\min}\\ 1 & \text{if}\ \ Qfc_i^t\geq Qfc_{i,\min}\end{cases} \tag{6.28}$$

Electrical Network The energy sold to the electrical network could be maximized or minimized according to the market penetration (MP).

$$\delta_i = \begin{cases} -1 & if \quad MP < MP_{threshold} \\ 1 & otherwise \end{cases} \tag{6.29}$$

2.4 Results and Discussion

The problem is here solved using mathematical programming techniques through a commercial optimization package (Lingo, www.lindosystems.org). In particular, the optimization problem is solved for a case study in Capo Vado site (Savona district) for a time period of one month (April 2009), which consists of a network of seven real refueling stations. Due to the lack of data regarding fossil fuel, electrical energy consumption, some territorial characteristics (distance, population density, etc.), meteorological measurements of wind speed and solar radiation, we applied the model to one point of production and seven existing refueling stations which we have their related data. In this section we assumed that this network of GHRS is the suitable one for the considered point of production. Table 6.1 reports the characteristics of the network of GHRS.

The developed approach supposes that the energy generated from the mixed renewable energy systems is used to ensure electrical energy and hydrogen fuel demands of the GHRS as well as a hydrogen market demand. The last is assumed to be constant and equals to 10 kg in time period. The hydrogen market could reflect different users that may buy hydrogen from the production plant.

The energy surplus is sold to the electrical network. It is assumed in this case study that the forecasted data from which the optimization problem must be solved are exactly equal to the historical data recorded in the Capo Vado site, and which consist of the daily wind speed and solar radiation, recorded at the height of 10 m. Fig. 6.4 shows the daily variation of the wind speed and solar radiation of Capo Vado site; it can be seen that the wind speeds range between 4.8 and 9.5 m/s. Generally, the wind speed takes an average value equal approximately to 7.54 m/s at 10 m of height. Whereas, solar radiation ranges between 0.46 and 7.3 kWh/m^2/day, with an average value of 4.8 kWh/m^2/day.

Ten wind turbines (ENERCON E-53) (www.enercon.de), with the following geometric and technical characteristics have been considered: $v_c = 2$ m/s, $v_r = 13$ m/s, $v_f = 25$ m/s, $P_r = 800$ kW, $H_{hub} = 75$ m. The power curve of the considered wind turbine supplied by the manufacturer is shown in Fig. 6.5. For the photovoltaic modules, it is assumed $S_{pv} = 10{,}000$ m^2, $\eta_{pv} = 0.11$, $\eta_{pc} = 0.86$, and $P_f = 0.9$.

In this work, the hydrogen demands are assumed to be known, and they are estimated using the real existing demands of the petrol stations available in the province. These real data are recorded from petrol stations of an Italian Petrol

TABLE 6.1 Characteristics of a network of GHRS

GHRS	Code	Longitude	Latitude	Distance (km)	Density of population (Inha/km^2)
Refueling station 1	PV 01363	44.4286	8.7587	52	7708
Refueling station 2	PV 01398	44.2682	8.4344	22	1750
Refueling station 3	PV 11348	44.3419	8.5438	33	1454
Refueling station 4	PV 01334	44.1688	8.3413	27	2160
Refueling station 5	PV 01377	44.0884	8.2072	45	1553
Refueling station 6	PV 01371	44.3211	8.4654	21	6168
Refueling station 7	PV 01351	44.1294	8.259	45	2277
Production point	–	44.3984	8.2193	0	–

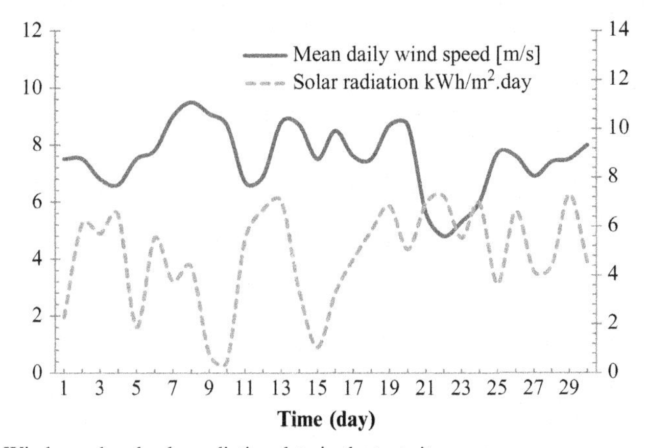

FIG. 6.4 Wind speed and solar radiation data in the test site.

Company. The demand of hydrogen used in this paper is calculated according to several scenarios of market penetration (5%, 15%, 25%). This hypothesis seems reasonable since in a short-term scenario, the demand of hydrogen cannot be equal to the one of petrol products consumed by vehicles. Fig. 6.6 displays

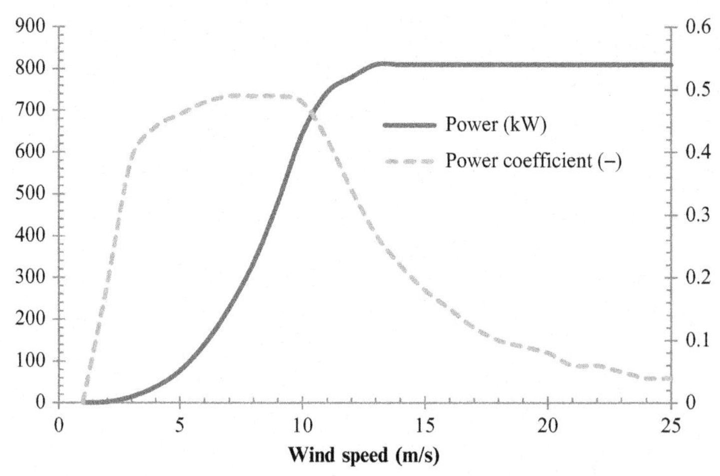

FIG. 6.5 Manufacturer power curve of the considered wind turbine.

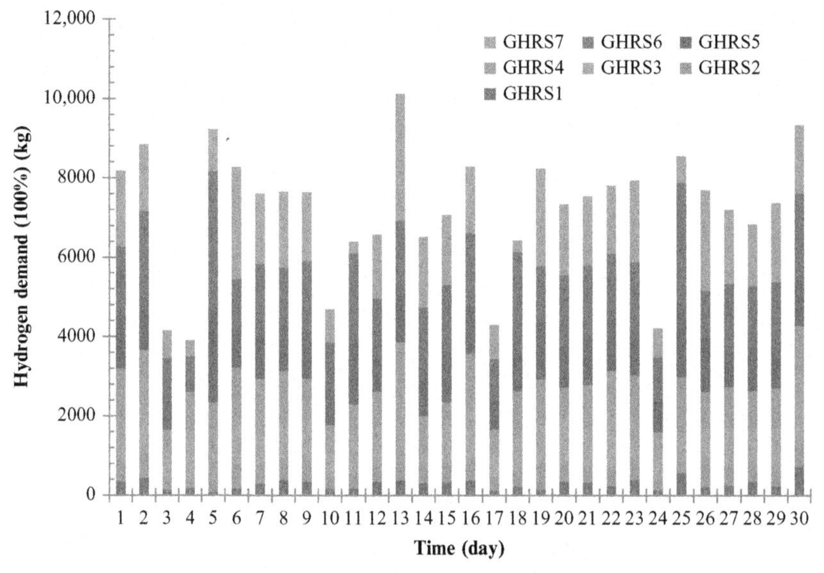

FIG. 6.6 Hydrogen demand of the network of GHRS.

the hydrogen demands of the network of the GHRS considered. It can be seen that the consumption behaviors of the considered stations in April month are different from each other, and it ranges between 54 and 3500kg.

Fig. 6.7 shows the daily electrical energy needs of the GHRS, they range between 1.7 and 318kWh. These demands are related to the real consumption recorded at petrol stations available in the province of Savona. These demands are almost coming from the electric network, but, since in this study, the GHRS

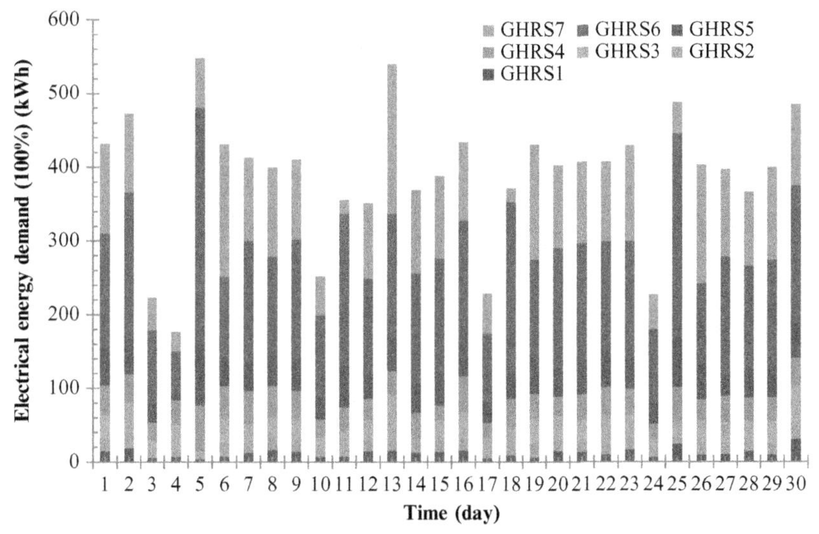

FIG. 6.7 Electrical energy demand of the network of GHRS.

is mentioned, this mean that electricity consumption needs to be satisfied by the energy produced by renewable energy system. These demands are mainly coming from lighting, pumps, vehicles machine washes, and others. The difference in electricity consumption among refueling stations is mainly due to the availability or not of special services (for instance, car washes machine) and to the demand of hydrogen for vehicles at each time instant that is directly linked to the pumps consumption. The electrical energy demands used in this case study are calculated according to the scenarios of the market penetration.

Fig. 6.8 displays the daily energy produced by the mixed renewable energy system; we should note that no distinction is considered between the energy produced by wind turbines and photovoltaic modules. It is assumed that it is a clean energy produced by a renewable energy system, since the economical aspect is not taken into account in this study. The daily produced energy ranges between 31 and 188 MWh. Fig. 6.9 shows the daily energy sent to the electrolyzer plant according to the three scenarios of market penetration in order to produce hydrogen. Fig. 6.10 reports the daily electrical energy surplus sold to the electrical network according to each scenario; it can be seen that the most part of electrical energy has been sold in scenarios where the market penetration equals to 5% followed by 15%, while in the third scenario (25%) the system did not exchange energy with the electrical network. This is due to the low demand of hydrogen with a high energy production in the two first scenarios and a high hydrogen demand in the third one. Fig. 6.11 displays the daily electrical energy supplied to the network of GHRS to satisfy their electrical energy needs including the energy consumed by the local hydrogen storage tanks.

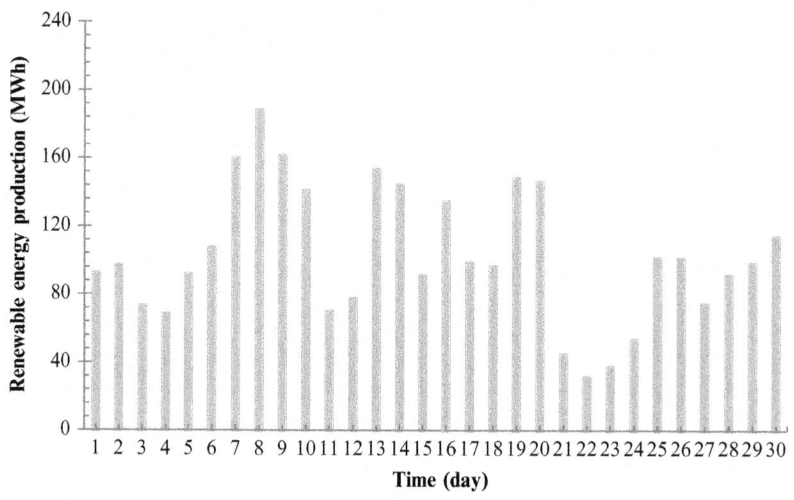

FIG. 6.8 Energy produced by the mixed energy system.

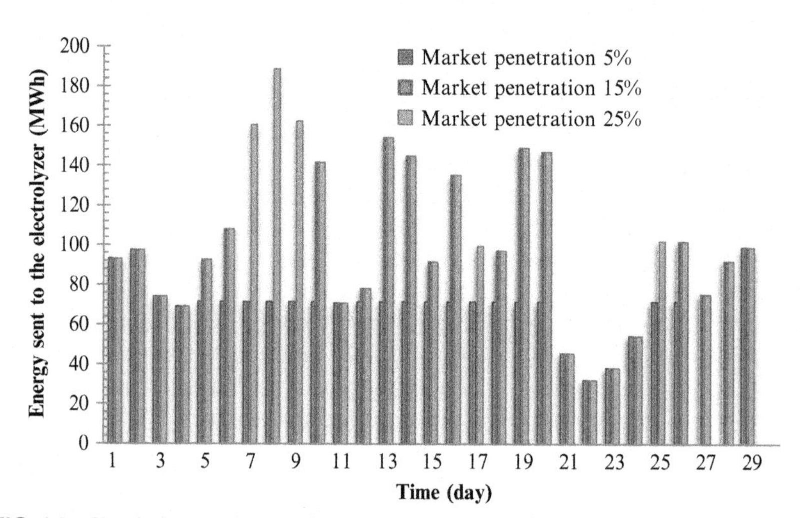

FIG. 6.9 Electrical energy sent to the electrolyzer plant.

Fig. 6.12 shows the trend of the energy stored along the month in the main hydrogen storage tank according to the scenarios of the market penetration. It appears that the higher amount of hydrogen stored is reached for a market penetration equals to 5% which is due to the low hydrogen demand. Figs. 6.13–6.19 show the level of hydrogen in each local storage system according to the three scenarios of the market penetration. It is observed that each GHRS exhibits a various storage tendency than the other which is mainly due to the difference in hydrogen demands. It appears from the figures that once the market

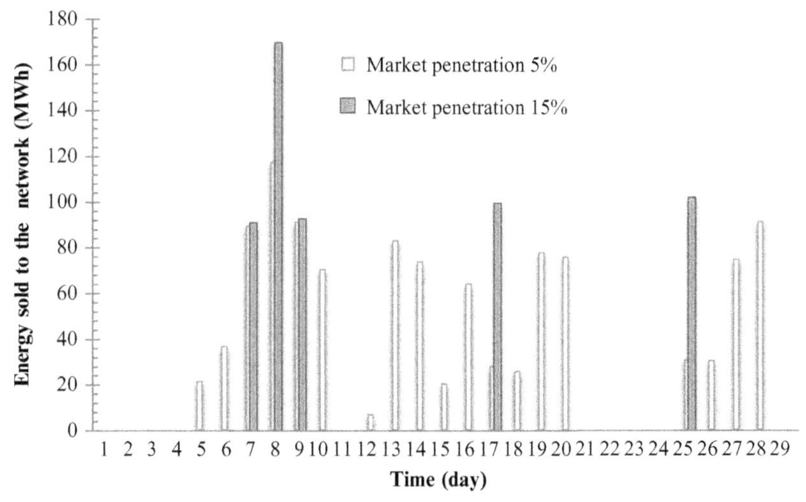

FIG. 6.10 Electrical energy sold to the network.

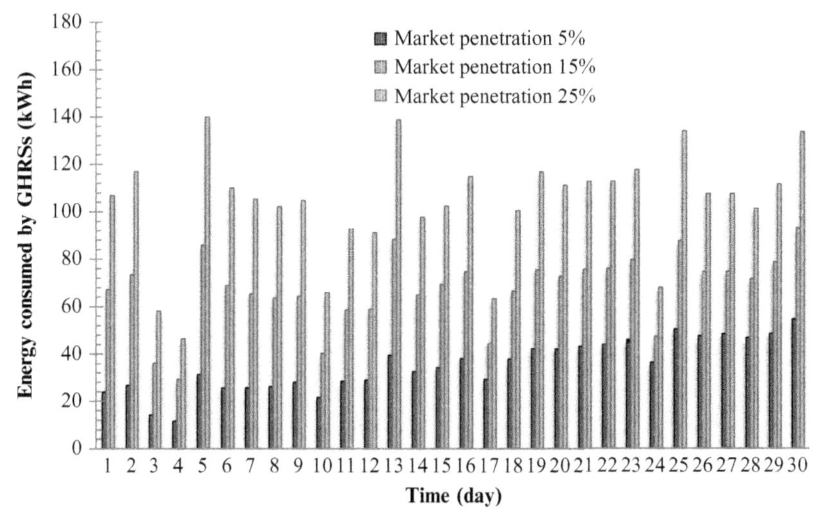

FIG. 6.11 Electrical energy consumed by GHRS.

penetration level of hydrogen increases the level of hydrogen stored in the hydrogen refueling stations as well as in the main storage system is increasing dramatically. This remark is mainly due to the augmentation of the hydrogen vehicles which necessitates more hydrogen demand. For the same market scenario, the change in the storage level is also dependent on the time of the week and on the size of the refueling station that serves hydrogen.

In general, all demands are always satisfied in each time period and for each scenario.

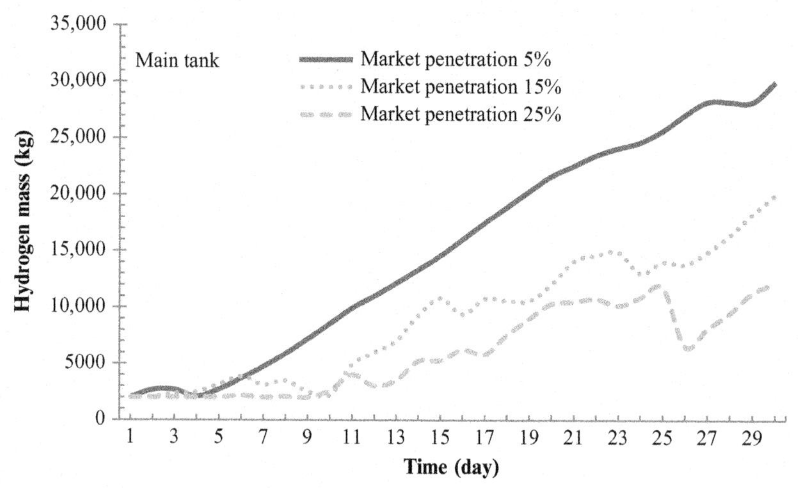

FIG. 6.12 Hydrogen storage level of the main tank.

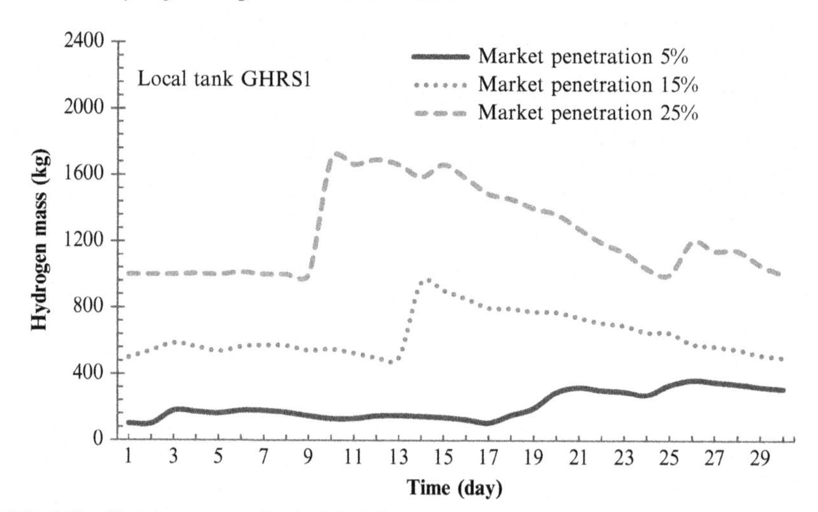

FIG. 6.13 Hydrogen storage level of GHRS1.

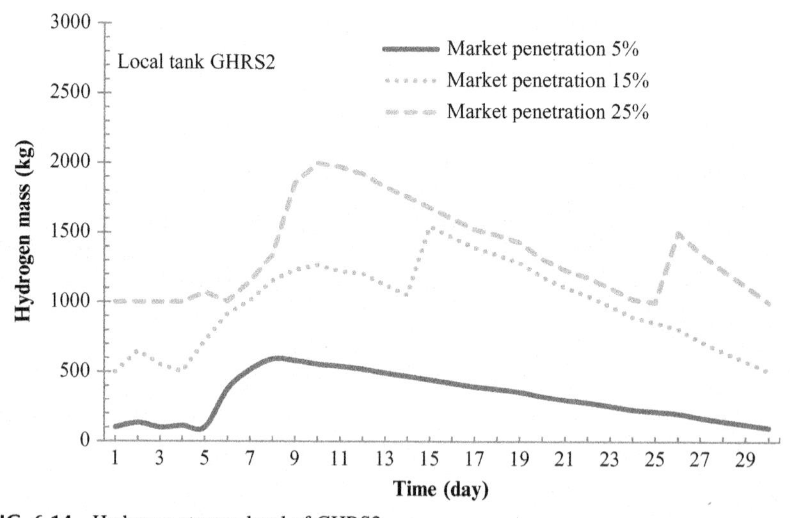

FIG. 6.14 Hydrogen storage level of GHRS2.

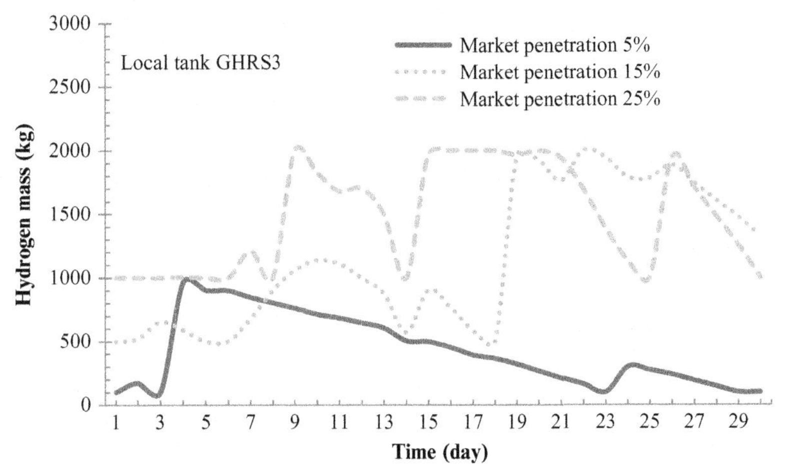

FIG. 6.15 Hydrogen storage level of GHRS3.

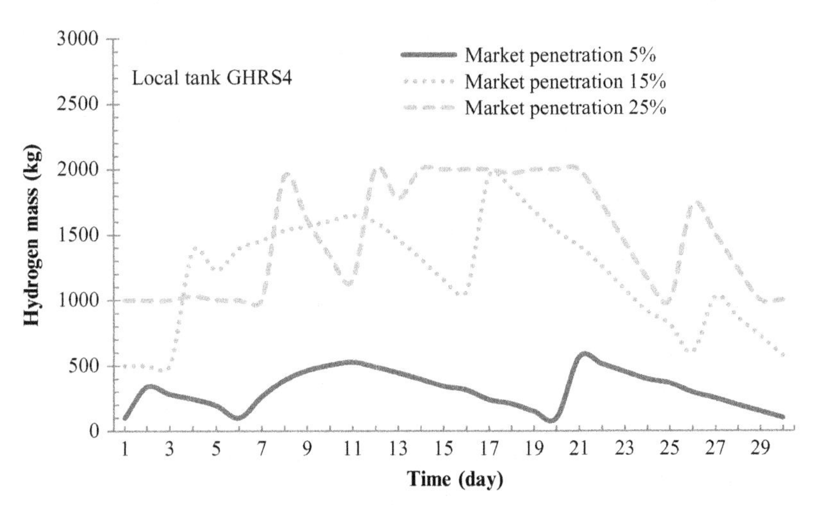

FIG. 6.16 Hydrogen storage level of GHRS4.

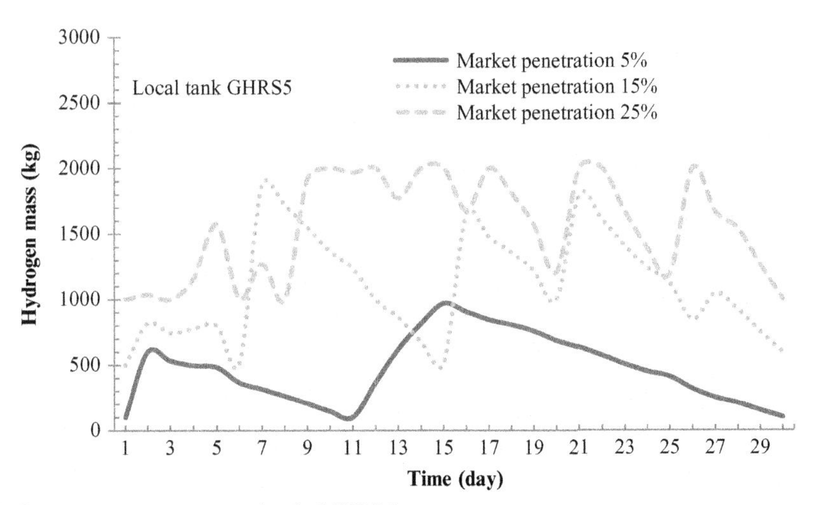

FIG. 6.17 Hydrogen storage level of GHRS5.

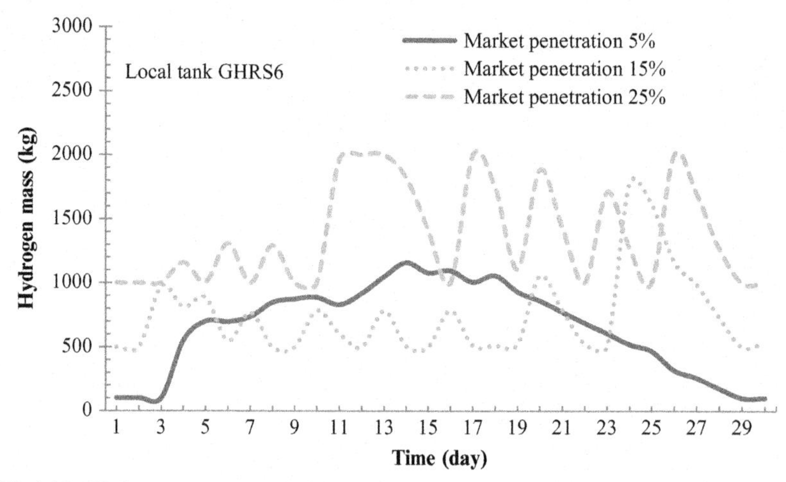

FIG. 6.18 Hydrogen storage level of GHRS6.

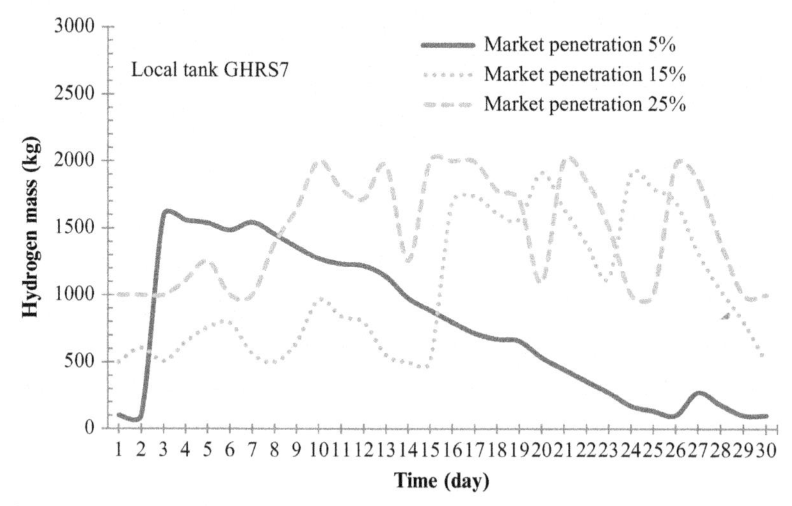

FIG. 6.19 Hydrogen storage level of GHRS7.

2.5 Conclusion

An optimization model of a network of GHRS is presented. The network of GHRS is completely powered by mixed renewable energy systems. The main decisions within the problem can be categorized into two types: first one related to optimal selection of the network of GHRS following roads network distance and population density criteria, while the second is the control and management of the hydrogen and electric flows that are sent to the GHRS. The optimization model has been simplified and then applied on one month basis to a case study in the province of Savona, Italy. Optimal results are reported taking into account the presence of an additional hydrogen industrial market and a connection with the electrical network. Results show that the demands are satisfied for each time period and for all market penetration scenarios. Future developments of the present work regard the possibility of introducing stochastic issues both in demand and resource predictions. In this case, in order to reduce the overall complexity of the problem, dynamic programming could be used. Then, the model could be detailed from the electrical and technological point of view. Moreover, an objective function based on hydrogen costs could also be adopted. Finally, the inclusion of the developed dynamic decision model in a decision support system could be helpful since it can allow to the user to interact with the mathematical models.

REFERENCES

Almansoori, A., Shah, N., 2006. Design and operation of future hydrogen supply chain snapshot model. Chem. Eng. Res. Des. 84, 423–438.

Almansoori, A., Shah, N., 2009a. Design and operation of a future hydrogen supply chain: multi-period model. Int. J. Hydrog. Energy 34, 7883–7897.

Almansoori, A., Shah, N., 2009b. Design and operation of a future hydrogen supply chain: multi-period model. Int. J. Hydrog. Energy 347883–347897.

Ball, M., Wietschel, M., Rentz, O., 2007. Integration of a hydrogen economy into the German energy system: an optimizing modelling approach. Int. J. Hydrog. Energy 32, 1355–1368.

Barbir, F., 2005. PEM electrolysis for production of hydrogen from RES. Sol. Energy 78, 661e9.

Benchrifa, R., Bennouna, A., Zejli, D., 2007. In: Role de l'hydrogene dans le stockage de l'electricite´ a base des Energies renouvelables.2IWH 2007; 27–29 October.

Bersani, C., Minciardi, R., Sacile, R., Trasforini, E., 2009. Network planning of fuelling service stations in a near-term competitive scenario of the hydrogen economy. Socio Econ. Plan. Sci. 43, 55–71.

Bokris, J., Veziroglu, T., 2007. Estimates of the price of hydrogen as a medium for wind and solar sources. Int. J. Hydrog. Energy 32 (12), 1605–1610.

Brey, J.J., Brey, R., Carazo, A.F., Contreras, I., Hernandez-Dıaz, A.G., Gallardo, V., 2006. Designing gradual transition to hydrogen economy in Spain. J. Power Sources 159, 1231–1240.

Campen, A., Mondal, K., Wiltowski, T., 2008. Separation of hydrogen from syngas using a regenerative system. Int. J. Hydrog. Energy 33, 332.

Clarke, R.E., Giddey, S., Ciacchi, F.T., Badwal, S.P.S., Paul, B., Andrews, J., 2009. Direct coupling of an electrolyser to a solar PV system for generating hydrogen. Int. J. Hydrog. Energy 34, 2531–2542.

Contaldi, M., Gracceva, F., Mattucci, A., 2008. Hydrogen perspectives in Italy: analysis of possible deployment scenarios. Int. J. Hydrog. Energy 33, 1630–1642.

Dagdougui, H., Minciardi, R., Ouammi, A., Robba, M., Sacile, R., 2010. A dynamic decision model for the real time control of hybrid renewable energy production systems. IEEE Syst. J. 4(3).

Demirbas, A., 2008. Present and future transportation fuels. Energy Sources Part A 30, 1473.

Doll, C., Wietschel, M., 2008. Externalities of the transport sector and the role of hydrogen in a sustainable transport vision. Energ Policy 36, 4069–4078.

Farrell, A.E., Keith, D.W., Corbett, J., 2003. A strategy for introducing hydrogen into transportation. Energ Policy 31, 1357–1367.

Greiner, C., Korpas, M., Holen, A.A., 2008. Norwegian case study on the production of hydrogen from wind power. Int. J. Hydrog. Energy 32, 1500–1507.

Guillén-Gosálbez, G., Mele, F., Grossmann, I., 2009. A bi-criterion optimization approach for the design and planning of hydrogen supply chains for vehicle use with economic and environmental concerns. AIChE J. 56, 650–667. https://doi.org/10.1002/aic.12024.

Hajimiragha, A., Fowler, M.W., Canilare, C.A., 2009. Hydrogen economy transition in Ontario-Canada considering the electricity grid constraints. Int. J. Hydrog. Energy 34, 5275–5293.

Hake, J.F., Linssen, J., Walbeck, M., 2006. Prospects for hydrogen in the German energy system. Energ Policy 34 (11), 1271–1283.

Honnery, D., Moriarty, P., 2009. Estimating global hydrogen production from wind. Int. J. Hydrog. Energy 34, 727–736.

Hugo, A., Rutter, P., Pistikopoulos, S., Amorelli, A., Zoiac, G., 2005. Hydrogen infrastructure strategic planning using multi-objective optimization. Int. J. Hydrog. Energy 30, 1523–1534.

Ingason, H.T., Ingolfsson, H.P., Pall, J., 2008. Optimizing site selection for hydrogen production in Iceland. Int. J. Hydrog. Energy 33, 3632–3643.

Joffe, D., Hart, D., Bauen, A., 2004. Modeling of hydrogen infrastructure for vehicle refueling in London. J. Power Sources, 13:13–22.

Johnson, N., Yang, C., Ogden, J., 2008. A GIS-based assessment of coal based hydrogen infrastructure deployment in the state of Ohio. Int. J. Hydrog. Energy 33, 5287–5303.

Jørgensen, C., Ropenus, S., 2008. Production price of hydrogen from grid connected electrolysis in a power market with high wind penetration. Int. J. Hydrog. Energy 33, 5335–5344.

Kamarudin, S.K., Daud, W.R.W., Yaakub, Z., Misron, Z., Anuar, W., Yusuf, N.N.A.N., 2009. Synthesis and optimization of future hydrogen energy infrastructure planning in Peninsular Malaysia. Int. J. Hydrog. Energy 34, 2077–2088.

Kim, J., Lee, Y., Moon, I., 2008. Optimization of a HSC under demand uncertainty. Int. J. Hydrog. Energy 33, 4715–4729.

Kim, J., Moon, I., 2008a. Strategic design of hydrogen infrastructure considering cost and safety using multiobjective optimization. Int. J. Hydrog. Energy 33, 5887–5896.

Kim, J., Moon, I., 2008b. The role of hydrogen in the road transportation sector for a sustainable energy system: a case study of Korea. Int. J. Hydrog. Energy 33, 7326–7337.

Konda, N.V.S.N., Shah, N., Brandon, N.P., 2011. Optimal transition towards a large-scale hydrogen infrastructure for the transport sector: the case for the Netherlands. Int. J. Hydrog. Energy 36, 4619–4635.

Kruger, P., Blakeley, J., Leaver, J., 2003. Potential in New Zealand for use of hydrogen as a transportation fuel. Int. J. Hydrog. Energy 28, 795–802.

Kruse, B., Grinna, S., Buch, C., 2002. Hydrogen Status og muligheter. Bellona rapport nr, vol. 6.

Kuby, M., Lines, L., Schultz, R., Xie, Z., Kim, J.G., Lim, S., 2009. Optimization of hydrogen stations in Florida using the flow-refuelling location model. Int. J. Hydrog. Energy 34, 6045–6064.

Lee, J.-Y., Yu, M.-S., Cha, K.-H., Lee, S.-Y., Lim, T.W., Hur, T., 2009. A study on the environmental aspects of hydrogen pathways in Korea. Int. J. Hydrog. Energy 34, 8455e67.

Li, Z., Gao, D., Chang, L., Liu, P., Pistikopoulos, E., 2008. Hydrogen infrastructure design and optimization: a case study of China. Int. J. Hydrog. Energy 33, 5275–5286.

Li, Y., Xue, B., He, X., 2009. Catalytic synthesis of ethylbenzene by alkylation of benzene with diethyl carbonate over HZSM-5. Catal. Commun. 10, 702.

Lin, Z., Chen, C., Ogden, J., Fan, Y., 2008. The least-cost hydrogen for Southern California. Int. J. Hydrog. Energy 33, 3009–3014.

Lin, Z., Ogden, J., Fan, Y., Chen, C., 2008. The fuel-travel-back approach to hydrogen station siting. Int. J. Hydrog. Energy 33, 3096–3101.

López, R., Bernal-Agustín, J.L., Mendoza, F., 2009. Design and economical analysis of hybrid PVewind systems connected to the grid for the intermittent production of hydrogen. Energ Policy 37, 3082–3095.

Martin, K.B., Grasman, S.E., 2009. An assessment of wind-hydrogen systems for light duty vehicles. Int. J. Hydrog. Energy 34, 6581–6588.

Melaina, M.W., 2003. Initiating hydrogen infrastructures: preliminary analysis of a sufficient number of initial hydrogen stations in the US. Int. J. Hydrog. Energy 28, 743–755.

Melendez, M., Milbrandt, A., 2008. Regional consumer hydrogen demand and optimal hydrogen refuelling station sitting. Technical Report NREL/TP-540e42224.

Nicholas, M.A., Handy, S.L., Sperlong, D., 2004. Using geographical information systems to evaluate siting and networks of hydrogen stations. Transp. Res. Rec. 1880, 126–134.

Ogden, J., 1999. Developing an infrastructure for hydrogen vehicles a Southern California case study. Int. J. Hydrog. Energy 24, 709–730.

Parker, N., Fan, Y., Ogden, J., 2009. From waste to hydrogen: an optimal design of energy production and distribution network. Transp. Res. Part E 46, 534–545.

Parker, N., Fan, Y., Ogden, J., 2010. From waste to hydrogen: an optimal design of energy production and distribution network. Transp. Res. E 46, 534e45.

Prince-Richard, S., Whalel, M., Djilali, N., 2005. A techno-economic analysis of decentralized electrolytic hydrogen production for fuel cell vehicles. Int. J. Hydrog. Energy 30, 1159–1179.

Sabio, N., Gadalla, M., Guillén-Gosálbez, G., Jiménez, L., 2010. Strategic planning with risk control of hydrogen supply chains for vehicle use under uncertainty in operating costs: a case study of Spain. Int. J. Hydrog. Energy 35, 6836e52.

Segura, I., Pérez-Navarro, A., Sánchez, C., Ibáñez, F., Payá, J., Bernal, E., 2007. Technical requirements for economical viability of electricity generation in stabilized wind parks. Int. J. Hydrog. Energy 32, 3811–3819.

Shane, S.R., Samuelsen, G.S., 2009. Demonstration of a novel assessment methodology for hydrogen infrastructure deployment. Int. J. Hydrog. Energy 34, 628–641.

Sherif, S., Barbir, F., Veziroglu, T., 2005. Wind energy and the hydrogen economy e review of the technology. Sol. Energy 18, 647–660.

Smit, R., Weeda, M., de Groot, A., 2007. Hydrogen infrastructure development in The Netherlands. Int. J. Hydrog. Energy 32, 1387–1395.

Stiller, C., Bunger, U., Møller-Holst, S., Svensson, A.M., Espegren, K.A., Nowak, M., 2009. Pathways to a hydrogen fuel infrastructure in Norway. Int. J. Hydrog. Energy 35, 2597–2601.

Strachan, N., Balta-Ozkan, N., Joffe, D., McGeevor, K., Hughes, N., 2009. Soft-linking energy systems and GIS models to investigate spatial hydrogen infrastructure development in a low carbon UK energy system. Int. J. Hydrog. Energy 34, 642–657.

Syste'mes solaires le journal de l'e´olien n_ 8 2011. Barometre eolien - eurobserv'er -Fevrier.

Thomas, C.E., Kuhnl, F., James, B.D., Lomax, F.D., Baum, G.N., 1998. Affordable hydrogen supply pathways for fuel cell vehicles. Int. J. Hydrog. Energy 23, 507–516.

Tzimas, E., Castello, P., Peteves, S., 2007. The evolution of size and cost of a hydrogen delivery infrastructure in Europe in the medium and long term. Int. J. Hydrog. Energy 32, 1369–1380.

Van den Heever, S., Grossmann, I., 2003. A strategy for the integration of production planning and reactive scheduling in the optimization of a hydrogen supply chain network. Comput. Chem. Eng. 27, 1813–1839.

Wietschel, M., Hasenauer, U., de Groot, A., 2006. Development of European hydrogen infrastructure scenarios—CO_2 reduction potential and infrastructure investment. Energ Policy 34, 1284–1298.

Chapter 7

Hydrogen Logistics: Safety and Risks Issues

1. INTRODUCTION

Hydrogen has been often recognized as the likely energy carrier for the future energy systems because it would represent the universal remedy for the growing concerns in accordance with fossil resource depletion, global warming, and increased air pollution. The benefits are motivated given the great number of primary energy sources used for its production, such as natural gas, coal, biomass, and water, contributing toward greater energy safety and flexibility (Hugo et al., 2005). Generally, hydrogen is produced, stored, and then transported to the end users; in general, it must be transported from production plants to the storage or demand points. So that, the delivery process of its supply chain brings new hazards exposures. Hence, a safe and sustainable transition to the use of hydrogen requires that the safety issues associated with the hydrogen have to be investigated and fully understood (Venetsanos et al., 2008). For planning and installing on a large-scale production and distribution infrastructure in urban areas, good standards and best practices are indispensable (Pasman and Rogers, 2010). Safe practices in the production, storage, distribution, and use of hydrogen are essential for the widespread acceptance of hydrogen technologies. Public authorities have to find the correct balance between socially important standards in terms of safety and promoting R&D in the hydrogen sector. It may be reasonable not to set standards too strictly during premarket development, in order to maintain enough stimuli for private investment in R&D (Zachmann et al., 2012).

Any failure in hydrogen systems could damage the public perception and decrease the ability of hydrogen to gain the worldwide approval. However, a good knowledge of these dangers and their consequences is intended to implement a safe design of systems using hydrogen. In these circumstances, it is possible to envisage the development of hydrogen as an energy carrier or fuel alternative with a low level of risk, enough to be individually and socially acceptable. It is important to investigate the risks of various parts of the distribution chain before any real implementation. The development of preventive and protective measures will require a reliable risk assessment methodology, which has to rely on awareness base of the hydrogen comportment in the logistics chain.

Hydrogen Infrastructure for Energy Applications. https://doi.org/10.1016/B978-0-12-812036-1.00007-X
© 2018 Elsevier Inc. All rights reserved.

2. HYDROGEN SAFETY PROPERTIES

2.1 Physicochemical Properties

The concept of hydrogen as a primary energy vector has received considerable attention, due to the environmental and energy security benefits compared to the conventional fossil fuels in terms of emissions and availability of supply. Hydrogen is the most abundant element in the earth. Many researchers recognize hydrogen as an excellent future energy alternative, playing both the role of a future fuel and the role of an energy carrier for electricity. In addition, the hydrogen has many advantages such as being the most elements that has high energy content per mass. Besides having good burning properties and high energy output per unit of mass, hydrogen has some drawbacks that may affect directly its safety and acceptability by the public. Hydrogen has physicochemical properties that are drastically different from traditional fuels. The key concerns are related to its low ignition energy, high flame speed, low flame visibility, colorless and odorless, and wide flammability range.

Table 7.1 presents the properties of hydrogen that are particularly relevant to safety.

There is a great interest in hydrogen flammability limits and its implications on fire safety and prevention in many applications including hydrogen applications and logistics. For all fuels, the hazard is due to the physical properties of the fuel—in this case due to the flammable and explosive nature of the fuel. In addition, hydrogen is a gas lighter than air, which means that it has the tendency to rise and diffuse rapidly when released into the atmosphere, dependent on the direction, rate, and pressure of the release. In addition, due to the low density, hydrogen can also diffuse rapidly through certain porous materials or systems that would normally be gas tight with respect to air or other gases. Hydrogen is highly flammable and very easily ignited.

TABLE 7.1 **Properties of hydrogen that are particularly relevant to safety (http://www.nrel.gov)**

	Hydrogen	Methane	Gasoline
Flammability limits (vol%)	4.0–75	5.3–15	1.0–7.6
Auto-ignition temperature (°C)	572	632	440
Ignition energy (mJ)	0.018	0.280	0.25
Deflagration index (bar m/s)	550	55	100–150
Limits of detonation in the air (vol%)	13–65	6.3–13.5	1.1–3.3
Coefficient of diffusion in the air (cm/s)	0.61	0.16	0.05
Max flame speed in the air (cm/s)	3.06	0.39	–

2.2 Hydrogen Safety Issues

A hazard is defined as a "chemical or physical condition that has the potential for causing damage to people, property, or the environment" (AIChE/CCPS, 2000). The primary hazard associated with any form of hydrogen is inadvertently producing a flammable or detonable mixture, leading to a fire or detonation. In general, a released quantity of hydrogen may be ignited immediately at the point of release, or it may be ignited after the cloud has been dispersed for a certain time, or it may not ignite at all. Release of hydrogen can be both instantaneous (for instance, the rupture of a compressor or buffer cylinder) or continuous (for instance, a leak in a pipe). In the event of a release of hydrogen, there are a number of potential risk scenarios such as:

(1) Dispersion of hydrogen (followed by ignition),
(2) Explosion of hydrogen (followed by a jet flame),
(3) Instant ignition with resultant jet flame.

2.2.1 Hydrogen Leak

The hydrogen gas contains the smallest molecule, which may increase the probability of leak through small holes and materials. In other words, due to its low viscosity, hydrogen is much more prone to leak from piping connections than other hydrocarbons. Hydrogen would leak approximately three times faster than natural gas and five times faster than propane on a volumetric basis (Rosyid, 2006). From other side, hydrogen benefits of lower density compared with other fuels, which in turn gives it a higher buoyant behavior in case of a failure in opening environment. Table 7.2 expresses the leak frequency occurrence for the case of hydrogen pipes as estimated by LaChance et al. (2009)). The hydrogen leak sizes were expressed in percent fractional flow area and grouped into five classes: $<0.01\%$, $<0.1\%$, $<1\%$, $<10\%$, and 100%. For instance, 0.1% means that the leak size is 0.1% of the cross sectional area of pipe. The generic leak frequencies apply to an average of all equipment of the type in question (in this case: pipe), in Table 7.2, the generic leak defines

TABLE 7.2 Leak frequency estimate for hydrogen pipes (LaChance et al., 2009)

Leak area	Generic leak frequencies	Hydrogen leak frequencies
0.01	7.8E−04	8.6E−06
0.1	1.0E−05	4.5E−06
1	4.0E−05	1.7E−06
10	5.4E−06	8.9E−07
100	5.3E−06	5.6E−07

all leaks that have occurred coming in particular from chemical, oil, compressed gas, and nuclear industries.

2.2.2 Hydrogen Ignition

Fires and explosions have occurred in various components of hydrogen systems as a result of a variety of ignition sources. Ignition sources have included mechanical sparks from rapidly closing valves; electrostatic discharges in ungrounded particulate filters; sparks from electrical equipment, welding, and cutting operations; catalyst particles; and lightning strikes near the vent stack. For instance, in case of an instant ignition of hydrogen, a jet flame can be produced.

2.2.3 Hydrogen Flash Fire

Depending on the pressure of release, weather conditions and the size of the smallest internal diameter of pipe work, hydrogen will disperse within its flammable range for several meters from the release. Flash fires can occur when a cloud of hydrogen in open air finds an ignition source. The flame is traveling back to the leak. It must be noted that a flash fire is a nonexplosive combustion of vapor cloud resulting from a release of hydrogen gas into the open air (Pasman and Rogers, 2010).

2.2.4 Jet Fire

The analysis of fires is particularly important because they have been found by many researchers as the cause of most frequent accidents. Although jet fires are often smaller than pool fires (in liquid hydrogen) or flash fires, they can also be very large, depending on the fuel discharge rate. Although jet fires have been studied by a number of authors, some of them can be found in Schefer et al. (2006a,b) and Houf and Schefer (2007). Jet fire is a turbulent diffusion of flame that results from the combustion of a flammable fuel continuously released. Flames originating from a leak are almost unidirectional.

2.2.5 Explosion of Hydrogen

Hydrogen explosion represents a considerable hazard. Hydrogen gas forms combustible or explosive mixtures with the atmospheric oxygen over a wide range of concentrations in the range 4.0%–75% and 18%–59%, respectively. Vapor cloud explosions involve a large release of hydrogen outdoors that mixes with air to form a large flammable cloud before ignition occurs (Pasman and Rogers, 2010). The strength of the explosion depends on the magnitude of confinement which in turn depends on the degree of confinement and could generate blast wave that can produce damages to the surrounding buildings and people (Baraldi et al., 2009). Many of the studies on hydrogen stations have dealt with explosion, deflagration, or detonation of hydrogen (Yamanaka et al., 2004; Fukuda et al., 2004; Xu et al., 2008).

3. RISK OF HYDROGEN SUPPLY CHAIN

It is important to investigate the risks of various parts of the supply chain, especially given the fact that hydrogen presents many peculiarities from the safety and risk viewpoint. The hydrogen system chain was assessed considering production, large-scale storage on the production site, refueling stations, and final utilization for automotive purposes (Landucci et al., 2010). The peculiarities of such infrastructure are mainly related to the spread of hydrogen storage installations in vulnerable areas, such as populated areas, commercial, and critical infrastructures. The installations might also be so close to customers such as hydrogen vehicles. Due to these facts, the widespread use of hydrogen requires the safety level to be at least not larger than those of existing fossil fuel technologies. Besides, the challenge for risk analysts is to treat many threads in a dynamical system, while most tools to ensure safety are designed to deal with individual plants and their components (Markert et al., 2017).

3.1 Risks in Hydrogen Systems

Like hydrocarbons, hydrogen as an activity related to energy generation has some safety implications that may start from the production side to the end uses (final service). The European Integrated Hydrogen project (EIHP) group claimed that the many ways in which hydrogen differs from conventional fuels make it necessary to perform detailed risk assessment for every stage in the hydrogen supply chain.

The use of hydrogen vehicles will require appropriate infrastructures for production, storage, and refueling stages. In particular, the storage problems can be considered as the most part involved in the supply chain where risks may arise. This is first due to the large amounts of hydrogen that are accumulated within the storage device, and second due to low hydrogen density, its low ignition temperature, and flammability, over a wide range of concentrations, which makes leaks a significant hazard for fire, especially in confined spaces (Casamirra et al., 2009).

Fig. 7.1 shows different parts of the hydrogen infrastructure involving risks. A common point can be illustrated from the figure is the involvement of hydrogen storage in different parts of the chain, for instance, hydrogen must be stored in transportation systems, at the refueling station, and onboard vehicles. Hence, great interest should be given to the storage side.

3.2 Risks in Hydrogen Refueling Stations

Hydrogen refueling stations must be as safe as gasoline stations. The facility of hydrogen fueling stations must be safe. As well we may know, the storage of hydrogen is an important aspect of fueling station design and construction. In fact, the storage system accomplishes two major roles in hydrogen

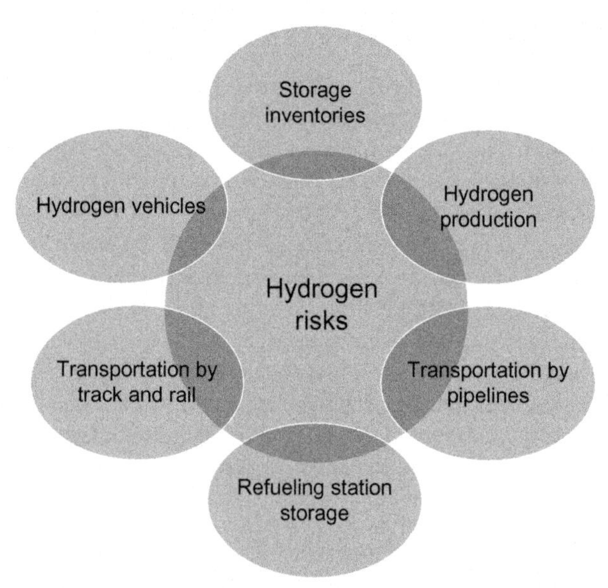

FIG. 7.1 Parts of the supply chain related to safety aspects.

delivery: increase of storage working capacity and regulation of delivery flow rate (Casamirra et al., 2009). The main safety aspects at the user interface are related to the risk associated with a potential ignition of a hydrogen leakage at the station or at the vehicle. Hydrogen refueling station may be a complex architecture since it must include additional devices that are essential to deliver the hydrogen to customers, such as compressor unit that is required to compress hydrogen to a required pressure, production facility (in case on-site refueling station). This complexity in addition to fuels properties may also give rise to the risk compared to the conventional RES. For these reasons, risk of hydrogen in the service station must be well evaluated and the code and standards for safety must be updated in order to take into account this hydrogen peculiarity. These safety issues and specifics may affect the public perception of installing a hydrogen refueling station, especially those that live close to the facility. Dayhim et al. (2014) implemented risk costs into a multiperiod optimization model with the objective function "minimization of the total daily social cost" of a hydrogen supply chain network. Zhiyong et al. (2011) presented a quantitative risk assessment study on gaseous hydrogen refueling station of 2010 World Expo. The Expo hydrogen station, located in the vicinity of the Expo site, is mainly used to fill fuel cell vehicles for 2010 World Expo. The main aim is to evaluate the risk of the station to personnel, refueling customers, and to third parties. The results have shown that the leaks from compressors and dispensers are the main risk contributors to first party and second party risks. This outcome leads to the conclusion that mitigation measures should be implemented in the first place on compressors and dispensers. Table 7.3 presents the probability of major accident that can cause one or more fatalities.

TABLE 7.3 Probability of a major accident causing one or more fatalities among customers (Zhiyong et al., 2011)

	Probability of accidents
Leak from compressors	1.20×10^{-3}
Leak from dispensers	1.17×10^{-3}
Pipe work-2 rupture	1.76×10^{-6}
Leak from vehicles fittings	9.52×10^{-6}
Others	$<10^{-6}$

Kikukawa et al. (2008) undertook a risk assessment of hydrogen fueling stations for 70-MPa fuel cell vehicles using data on hydrogen behavior at 70 MPa which was extrapolated from existing 35-MPa hydrogen stations data. The study results enable the identification of the safety issues that must be resolved to maintain safety of the hydrogen refueling station that operates on high pressure such as 70 MPa. Results of the study suggested that the safety distance could be maintained to 6 m for 70-MPa hydrogen stations as the 35-MPa ones.

Same authors, in Kikukawa et al. (2009), studied risk on liquid hydrogen refueling stations. They assumed two forms of explosion that might happen at liquid hydrogen fueling stations. One is a diffusion explosion where leaked liquid hydrogen vaporizes and mixes with the air and ignites resulting in a diffusion explosion, and the other is a premixed explosion where leaked liquid hydrogen remains on the residual area and mixes with the air and ignites resulting in a premixed explosion. In addition, same authors have highlighted two consequence levels, namely, blast pressure and flame. As regards the first kind of consequences, they result from a boiling liquid expanding vapor explosion (BLEVE) or vapor confined explosion (VCE). As regards the jet flame, results have shown that 14-mm wide hole produced a 10-m long jet flame, a 1.0-mm wide hole produced a 1.7-m long jet flame, and a 0.2-mm wide hole produced no jet flame. Work presented by Houf and Schefer (2007) describes an application of quantitative risk assessment methods to help establish one key code requirement: the minimum separation distances between a hydrogen refueling station and other facilities and the public at large. The separation distances were calculated using a Sandia developed model for predicting the radiant heat fluxes and flammability envelopes from high pressure releases of hydrogen (Houf and Schefer, 2007). Fig. 7.2 provides an example of deterministic separation distances based on one possible consequence of a hydrogen leakage event: the radiant heat flux from an ignited hydrogen jet. It shows the separation distances required to limit the exposure of a person to a radiant heat flux of 1.6 kW/m^2 which is generally accepted as a level that will not result in harm to an individual even for long exposures.

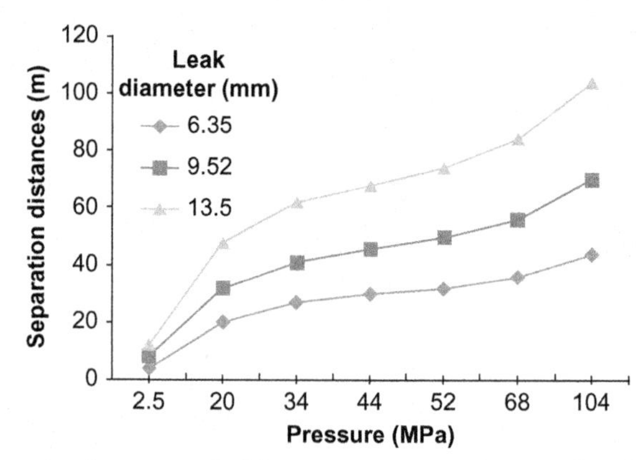

FIG. 7.2 Separation distances required for an exposure to a radiant heat flux of $1.6\,kW/m^2$ generated by a jet fire (LaChance et al., 2009).

4. METHODOLOGY FOR HYDROGEN RISKS ANALYSIS

4.1 Risk Definition

Risk is "a measure of human injury, environmental damage or economic loss in terms of both the incident likelihood and the magnitude of the loss or injury" (AIChE/CCPS, 2000). Also, according to the same reference (AIChE/CCPS, 2000), risk assessment is "the process by which the results of a risk analysis are used to make decisions."

Risk has been considered as the chance that someone or something that is valuated will be adversely affected by the hazard (Woodruff, 2005) while "hazard" is any unsafe condition or potential source of an undesirable event with potential for harm or damage (Reniers et al., 2005). Moreover, risk has been defined as a measure under uncertainty of the severity of a hazard (Høj and Kröger, 2002), or a measure of the probability and severity of adverse effects (Haimes, 2009). Risk uncertainty has always been a problem in risk assessment studies.

4.2 Risk Analysis

Risk analysis is defined as "the development of a quantitative estimate of risk based on engineering valuation and mathematical techniques for combining estimates of incident consequences and frequencies" (AIChE/CCPS, 2000). Proceeding by risk analysis approach, two values of risk have to be measured, namely, the magnitude of the hazard and the probability that the hazard will occur. It is a very crucial phase of risk management since its knowledge provides the first stage of handling hydrogen risks. Fig. 7.3 displays different risk analysis approaches available.

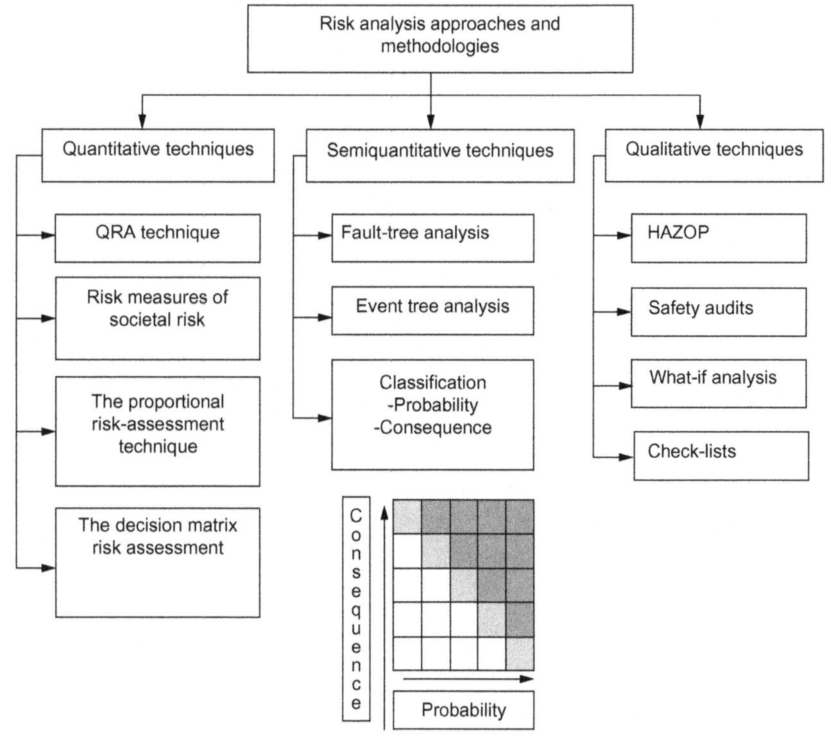

FIG. 7.3 Risk analysis approaches and methodologies.

4.3 Quantitative Risk Assessment Technique

The QRA requires the evaluation of the hazard and its consequences for each hydrogen release scenario that may occur in different nodes of the hydrogen supply chain as can be seen from Fig. 7.4. It is often used to quantify the risk around hydrogen facilities and support the communication with authorities during the permitting process. According to Marhavilas et al. (2011), this tool offers the possibility to define four types of objects: unprotected people, cars, residential, and buildings. Based on the findings from the QRA, potential measures to control and/or reduce the risk can be suggested and the effect of the measures evaluated (Haugom and Friis-Hansen, 2010).

Quantitative risk analysis can be also performed following the use of decision matrix risk assessment. Actually, this method consists of the estimation of risk based on the probability and the consequence of an event (Ayyub, 2003; Marhavilas and Koulouriotis, 2008). The product of consequence/severity and likelihood of a given impact measures the risk. Then, the risk of such an impact is quantified (Fig. 7.5). The risk is then compiled to a format so that it can be compared with the risk acceptance criteria applicable for each of the specific hydrogen infrastructure.

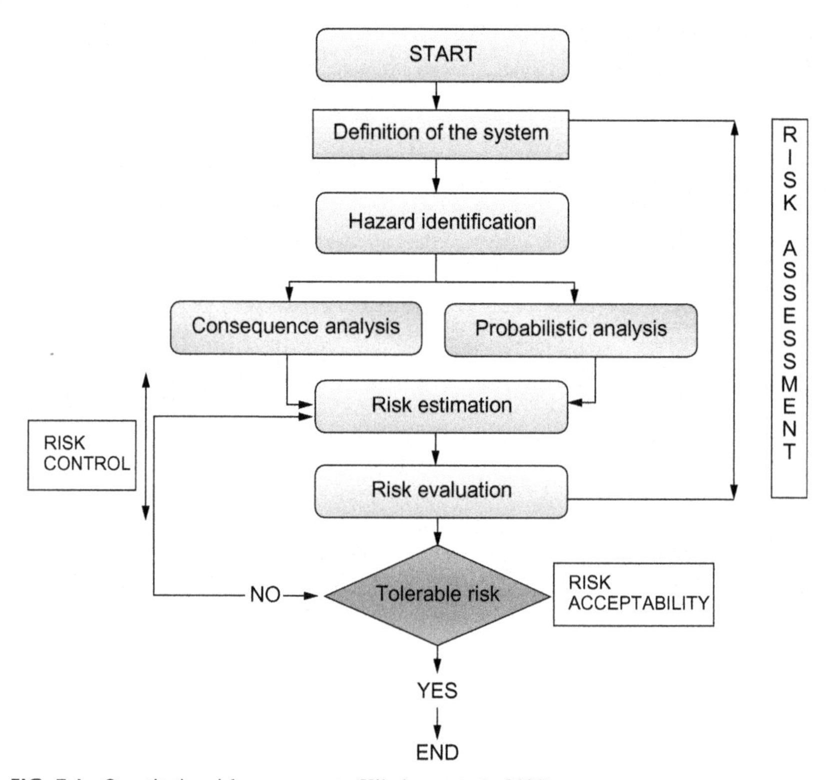

FIG. 7.4 Quantitative risk assessments (Kikukawa et al., 2009).

		Probability level			
		A	B	C	D
Consequence severity level	I	8	1	12	29
	II	0	0	0	6
	III	8	1	10	8
	IV	2	0	4	14
	V	00	1	5	13

FIG. 7.5 Risk map in case of liquid hydrogen refueling station (Kikukawa et al., 2009).

4.4 Hazard Identification

Identifying hazards is fundamental for ensuring the safe design and operation of a system in process plants and other facilities (Dunjó et al., 2010). Among hazard identification techniques, HAZOP (Hazard and Operability Studies) is one of the methods that are widely considered to be effective to identify the source of a risk; also it performs well for safety assessment in the process industry. In fact, HAZOP identify how a complex system can fail and to determine qualitatively whether the process design is robust and whether the existing safeguards are adequate (HyApproval WP2, 2008). In the context of hydrogen, analyzing the literature, it is worth to note that one main knowledge gap in the hazard identification stage is the current lack of clarity about the main components of a hydrogen delivery and storage infrastructure and their detailed design. This lack of data makes the operation of hazard identification quite complex.

4.5 Historical Analysis of Hydrogen Accidents

Historical analysis of the accidents occurring in the past is required, which may be helpful to determine and focus on the ones that occurs, the most probable and the ones that lead to high consequences. Many data can be retrieved by the most famous website dedicated to hydrogen entitled "H2 Tool and Lesson Learned" accessible by https://h2tools.org/ (last access July 2017). The American database is being developed by the US Department of Energy and it is a database-driven website intended to facilitate the sharing of lessons learned and other relevant information gained from actual experiences using and working with hydrogen. The database also serves as a voluntary reporting tool for capturing records of events involving either hydrogen or hydrogen-related technologies. The focus of the database is on characterization of hydrogen-related incidents and near-misses, and ensuing lessons learned from those events. The h2incidents database is organized enabling the access to different reported hydrogen incidents, thus regarding different viewpoints, namely, the probable causes of incidents, their contributing factors, the damages and injuries that result, and equipment involved in the accidents. Another database which contains a database of historical analysis of hydrogen is the MHIDAS (major hazard incident accident data service) (MHIDAS, 2008). It is a database which includes all accidents involving hazardous materials that have off-site impact, and also those which only have the potential to lead to an off-site impact. Such impacts include human casualties or damage to plant, property, or the natural environment. The analysis allowed us to determine the historically most diffused causes and related consequences that might happen when hydrogen is used in any kind of process (transport, industrial, process, etc.). According to Gerboni and Salvador (2009), over a total of 118 records were related to accidents involving hydrogen.

According to Fig. 7.6 published by the same authors, the most frequent cause for an accident was identified as the mechanical failure, although an equivalent share of events had not their cause explicitly determined.

From the hydrogen accidents viewpoint, major knowledge gaps exists in analyzing the accidents, thus specially is first due to the lack of handling hydrogen since it is a new energy for future, not yet used by public and other entities

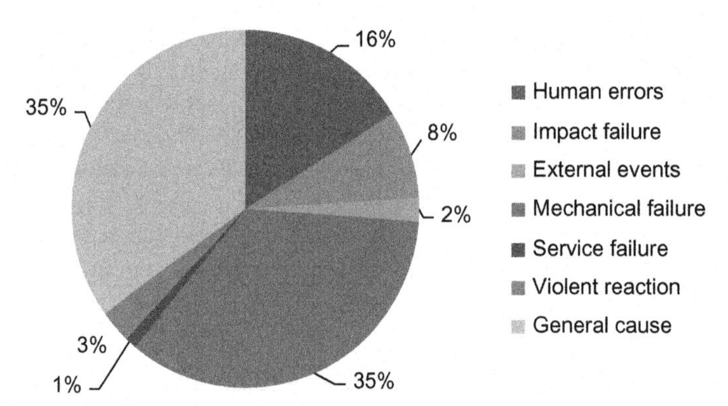

FIG. 7.6 Shares of total possible causes for accidents involving hydrogen (Gerboni and Salvador, 2009).

and second to the insufficient of historical data, since no mature sufficient experience is reached yet.

4.6 Event Tree Analysis

Event trees are commonly used to keep track of cause-effect relationships, and system and material behavior characteristics (Gerboni and Salvador, 2009). Event tree analysis (ETA) is a graphic technique that uses decision trees and logically develops visual models of the possible outcomes of an initiating event (Marhavilas et al., 2011). The ETA technique may be applicable to the design, construction, and operation stages of a hydrogen-related system.

4.6.1 ETA for Hydrogen Release: Pipeline

The failure rate of a pipeline has units of the number of failures per year per unit length of the pipeline, 1/yr km, assuming uniform conditions along the pipeline section of interest. It is somewhat different from the case of a point source of failure in which the rate is defined as the number of failures per year (Jo and Crowl, 2008).

Fig. 7.7 presents the event tree for hydrogen pipeline transmission. Initiation can be realized with a very-low-energy ignition source, i.e., 0.02 MJ (Steven and John, 1990). It is worth to note from Fig. 7.7 that the possibility of a significant flash fire or vapor cloud explosion resulting from far delayed remote ignition is extremely low due to the buoyant nature of the hydrogen. But within few seconds after the start of the release, a large flammable gas cloud could be formed due to the turbulent mixing between hydrogen and ambient air. This cloud has the potential to produce an explosion due to the nature of hydrogen gas. If the released gas ignited immediately with the rupture of pipeline, it makes a jet fire just after a short-lived fireball.

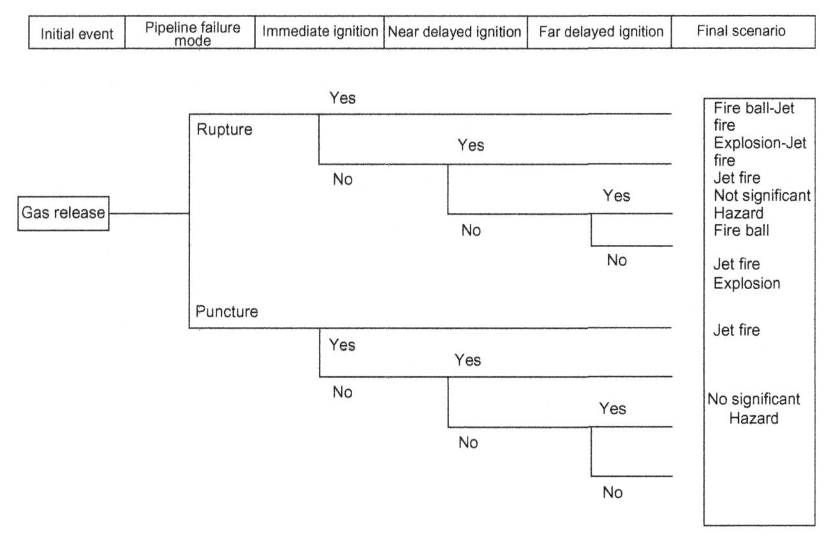

FIG. 7.7 Event tree for pipeline hydrogen transmission (Jo and Ahn, 2006).

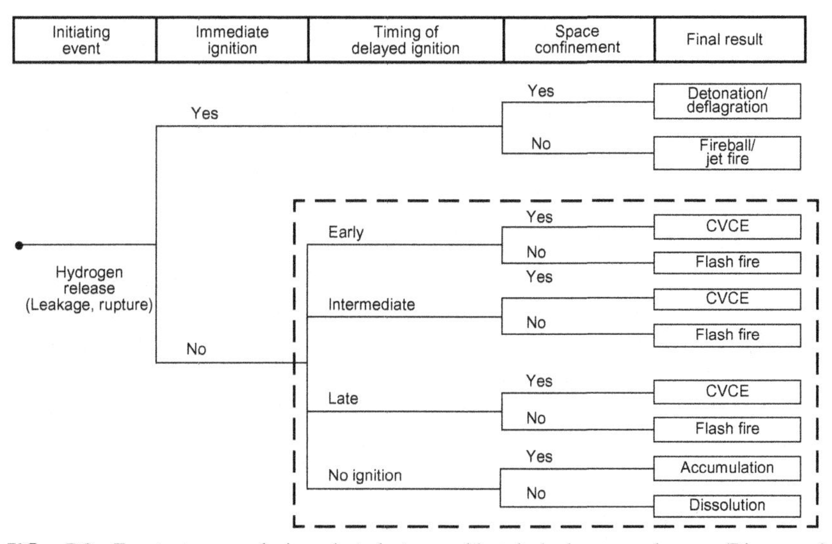

FIG. 7.8 Event tree analysis adapted to accidental hydrogen releases (Rigas and Sklavounos, 2005).

4.6.2 *ETA for Hydrogen Release: Storage Tank*

The event tree analysis regarding accidental hydrogen release is displayed in Fig. 7.8. It appears that the critical factors that may affect substantially the final outcome are two: the time of ignition of the resulting cloud and the confinement where the failure occurs.

It is obvious that, unless an immediate ignition takes place, there is some time of dispersion that intervenes between release and ignition. If real time hydrogen flammability zones were known, it would be possible for preventive measures to be taken and emergency response planning to be prepared against fires and explosions (Rigas and Sklavounos, 2005). Consequently, major issue arises regarding the computation of the dispersion succeeding an accidental hydrogen release.

Hydrogen dispersion may be considered "safe" only when no ignition occurs and no space confinement exists or well ventilation exists.

4.7 Consequence Calculation

According to the literature viewpoint, different methods and tools have been used to study the consequences of hydrogen accidents related to specific hydrogen economy infrastructure. These methods could be divided into three approaches:

1. The use of mathematical modeling
2. The use of experimental studies
3. The use of software and computational fluid dynamics (CFD)

We note that the three methods introduced previously have their own advantages and drawbacks, and they may provide a variety of knowledge as regards the comportment and how it behaves with the hydrogen substance once it is released from facility, thus, under different circumstances.

4.7.1 Mathematical Modeling

This method is based upon analytical method that describes the release of hydrogen. The paper presented by Houf and Schefer (2007) presented results from models for the radiative heat transfer from hydrogen jet flames and the concentration decay of unignited hydrogen jets for unintended release events involving high-pressure gas storage systems and fuel dispensers. The models are based on a combination of empirical correlations and analytical models, as well as a numerical model of the temporal blowdown of a hydrogen storage tank.

4.7.2 Experimental Studies

Many experimental studies have been done to highlight the hazards related to hydrogen. The aims of these works were to:

Test the usefulness of correlations for predicting flame length & radiant heat flux Measurement of the dimensional and radiative properties of large-scale vertical hydrogen jet Study the parameters that influences the hazard, namely, leak diameter, pressure.

Measurements done by Schefer et al. (2007) were performed to characterize the dimensional and radiative properties of large-scale vertical hydrogen jet flames. Thus measurements were obtained at storage pressures up to 413 bar (6000 psi) in this study. It was found that the flame length results show that lower pressure engineering correlations based on the Froude number and a nondimensional flame length also apply to releases from storage vessels at pressures up to 413 bar (6000 psi). Similarly, radiative heat flux characteristics of these high-pressure jet flames obey scaling laws developed for low-pressure, smaller scale flames and a wide variety of fuels. The same authors have performed in large-scale vertical flames to characterize the dimensional, thermal, and radiative properties of an ignited hydrogen jet. These data are relevant to the safety scenario of a sudden leak in a high-pressure hydrogen containment vessel. These data are relevant to the safety scenario of a sudden leak in a high-pressure hydrogen containment vessel.

4.7.3 Software and Computational Fluid Dynamics (CFD)

It is well known that it is quite expensive to undertake experimental real hydrogen release and combustion in real-scale configurations, the use of computational fluid dynamics (CFD) modeling for safety purposes is increasing in this field. The CFD modeling also permits the investigation of releases in real-world environments incorporating the environment conditions and using specific structures of infrastructures. Wilkening and Baraldi (2007) have studied an accidental release from a large pipeline for methane and hydrogen under similar conditions in a numerical viewpoint, thus for the aim to compare hydrogen pipelines to natural gas pipelines whose use is well established today. Within their paper, they compared safety implications in accidental situations. As a tool of consequence modeling, the CFD tool has been used to investigate different properties such as density, defeasibility, and flammability limits of hydrogen and methane on dispersion process. Since it is an accidental release of the substance, release is simulated through assimilating a hole between the high pressure pipeline and its environment. The numerical simulation has been made in the presence and the absence of the wind so to include its effect as regards the dispersion of the gases.

In the design of hydrogen supply chain, besides taking into account the economic and environmental aspects, the risk management is also an important side that must be given careful attention. There are two different perspectives to evaluate the hydrogen risk: (1) risk of hydrogen facilities to the population and environment, and (2) risk of hydrogen developing hydrogen projects. In the first definition, risk is presented in the previous sections of this chapter taking into account the physical properties of hydrogen. The second definition of risk is in turn related to the financial risks of developing hydrogen-related projects.

In fact, hydrogen projects are surrounded by many uncertainties which may lead to failures to complete the project or in meeting its financial goals. Hence it is important to identify the most economic routes and evaluate impacts of various scenarios of developing a supply infrastructure. The project developers may have many attitudes toward the financial risk associated with the investment on a project under uncertainty, for instance, risk adverse aims to avoid the unfavorable conditions focusing on solutions with lower variability for a certain budget.

Many solutions are proposed to avoid the financial risks of hydrogen projects using:

- The financial tools, which may include insurance and reinsurance policies, alternative risk transfer instruments, contingent capital, and credit enhancement products.
- Understand and prioritize the financial risk.
- Identify and understand scenarios that may lead to high risks.

In the literature, the component of financial risk is not well studied; there is a lack of research studies that investigate the risk associated hydrogen projects. Sabio et al. (2010) provided a mathematical programming for long-term design and planning of hydrogen supply chains for vehicle use under uncertainty in their economic performance with the ability to handle the financial risk associated to market changes. In their paper, authors have explicitly measured the financial risks via the worst case, which is associated to the objective function as an additional criterion to be optimized.

4.8 Risk Acceptance Criteria

It was accepted by the IEA hydrogen safety group that is paramount to determine appropriate risk acceptance criteria that ensure acceptable safety levels for the emerging hydrogen fueling infrastructure. According to EIHP (Haugom, DNV), there are several alternative strategies for developing risk acceptance criteria, such that:

- Comparing with statistics from existing petrol stations, giving an historical average risk level
- Comparing with estimated risk levels from risk analyses
- Comparing with general risks in society.

For instance, the risk acceptance criteria can be that to ensure that the risk level associated with hydrogen systems is similar to or smaller than the risks associated with comparable existing nonhydrogen systems that are generally worldwide accepted in society.

The main important methodology used for risk acceptance criteria is the "ALARP" criteria proposed also in the Final Report "Harmonised Risk Acceptance Criteria for Transport of Dangerous Goods" by European Commission DG-MOVE. The ALARP criteria (As Low As Reasonable Practicable) consisting of either qualitative or cost-benefit criteria for evaluation of additional

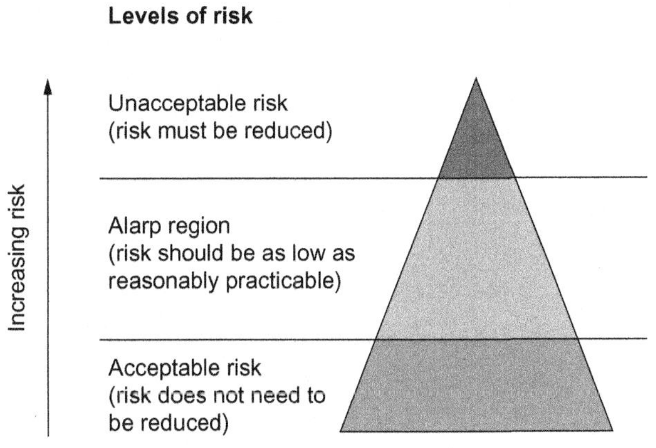

FIG. 7.9 Alarp region (DG MOVE, Report No: PP070679/4).

restrictions or safety measures. This aims to ensure that safety measures are optimized, taking account of the costs and benefits of risk reduction (Fig. 7.9).

The following figure shows the societal risk curve based on the ALARP principle.

The upper line in Fig. 7.10 represents the risk acceptance curve. The region between this line and the lower line denotes the ALARP region (As Low As Reasonable Practical). For scenarios with risk levels (that lie) between these lines the risk should be reduced if practical, typically subject to cost-benefit analysis. For scenarios with risk levels above the upper curve, measures to reduce the risk must be implemented.

According to risk accepted adopted by the EIHP (European integrated hydrogen platform), three main acceptable levels of risk could be attributed to the infrastructure related to hydrogen refueling stations. These levels differ according to type of party that is exposed; EIHP categorized these into three parties, namely, first party risk, second party risk, and third party risk. For the first party risks, the individual probability of fatality caused by hydrogen process-related events on the refueling station should not exceed 10^{-4} per year. For the second party risks, the probability of a major accident causing one or more fatalities among customers shall not exceed 10^{-4} per year. For the third party risks, both individual and societal risk measures should be considered (e.g., risk contours and FN curves) (EIHP2, Zhiyong et al., 2011).

4.9 Hydrogen Infrastructure Regulations

Hydrogen has unique physical and chemical properties which present benefits and challenges to its successful widespread as a fuel. Hydrogen can be used as safely as other common fuels we use today when guidelines are observed and

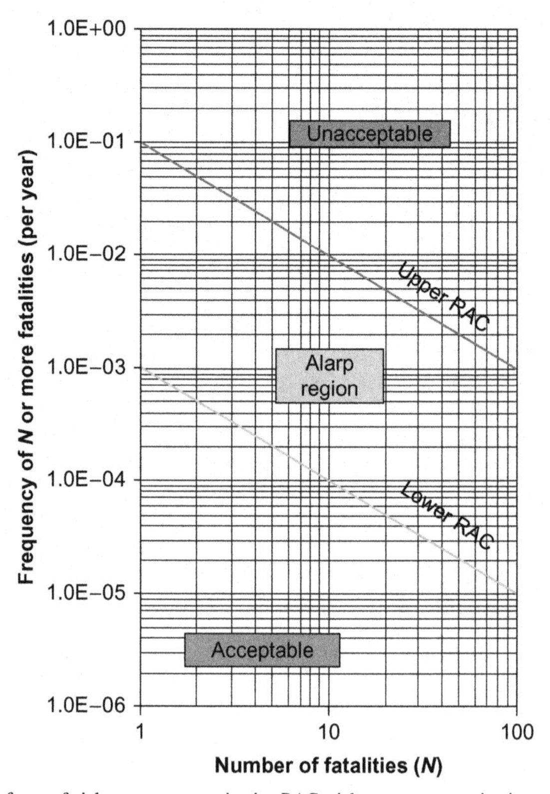

FIG. 7.10 FN form of risk acceptance criteria. *RAC*, risk acceptance criteria.

users understand its behavior. For this purpose, hydrogen regulations must be developed to provide the information needed to build, maintain, and operate hydrogen facilities including hydrogen fuel cell vehicles. A catastrophic failure in any hydrogen involved project could damage the public's perception of hydrogen and fuel cells. The safety hydrogen regulations should be developed including the production, storage, distribution, and the final use of hydrogen. They are essential for the widespread acceptance of hydrogen by government and public.

In the United States, Sandia National laboratories with the Department of Energy is the leader in developing technical research papers regarding the safety issues of hydrogen. They developed a technical basis for assessing the safety of hydrogen-based systems for use in the development/modification of relevant codes and standards. Two databases have been made available for public to share information related to hydrogen incidents and lessons learned and safety practices, design, and operation (https://h2tools.org/bestpractices).

In October 2014, the European Parliament emanated the Directive 2014/94/EU. The aim of the Directive is to build up refueling points for alternative fuels (electricity, liquid natural gas, compressed natural gas, hydrogen) in the EU, with common standards for their design and use. In Italy, the Legislative Decree of December 16, 2016, no. 257 for the application of the Directive 2014/94/EU of the European Parliament has been emanated. It entered into force from 14/01/2017 (http://www.h2it.org/2017/decreto-legislativo-16-dicembre-2016-n-257).

The regulations aim to guaranty the following objectives:

- minimize the cause of hydrogen release, fire, and explosion
- limit, in case of incident event, damage to person
- limit, in case of incident event, damage to buildings and adjacent structures
- allow to rescuers to operate in security conditions

The Italian Working Group on the hydrogen fire prevention safety issues (Grasso et al., 2005) has considered that three main topics related to hydrogen infrastructure need to be regulated (Italian ministry of the interior):

a. the hydrogen refueling stations
b. the hydrogen vehicle components
c. the hydrogen transport in pipelines

The same Italian Working Group presented theoretical and experimental results that are carried out on the hydrogen fire prevention safety issues in the field of the hydrogen transport in pipelines. According to the Group, the published current work is concluded to be considered as a document to begin discussions with the Italian stakeholders that are involved in hydrogen applications. In particular, from one hand, the theoretical framework developed aims at issuing a draft document that is based on Italian regulations in force on the natural gas pipelines. In fact, these latter have been reviewed, corrected, and integrated with instructions suitable to the use of hydrogen gas in pipelines. On the other hand, the experimental component has been designed and installed at the University of Pisa, an apparatus which will be considered as the simulation platform of hydrogen releases from pipelines. Another project entitled "The Zero Regio" project includes two demonstration projects, one in Germany (Frankfurt, region of Rhein-Main) and one in Italy (Mantova-Valdaro, region of Lombardy). The demonstrator projects are related to hydrogen refueling stations. The map on the Mobilitah2 page (http://www.mobilitah2.it/stations) shows the existing refueling points as of 2016 and a hypothesis of distribution between the various Italian cities of the planned stations expected to enter into operation by 2020 and by 2025, for both passenger cars and buses. The regulation requires safety distances of 50% larger than those mentioned in the draft "technical rule" which was applied in the design of the ENI (Italian biggest oil company) station. A 3×2 m concrete wall became safety requirements (Backhaus and Bunzeck, 2010).

5. CONCLUSION

Hydrogen has a long history of safety use in the chemical, manufacturing, and utility industries; however, as a large and widespread scale where general public can handle hydrogen infrastructure, it may create safety issues. For this main reason, the safety technological conditions for the use of hydrogen as an energy carrier must be well known and deeply studied in order to deliver appropriate safety codes and standard to minimize risks on public. The chapter discusses the risks of using the hydrogen in different components of its logistic chain as well as it discusses about the safety issues related to hydrogen. A detailed literature review is summarized which contains the state of art of hydrogen risks and the methodological approach followed by the worldwide scientific and technical communities to evaluate risks that results from hydrogen accident.

REFERENCES

AIChE/CCPS, 2000. Guidelines for Chemical Process Quantitative Risk Analysis, second ed. American Institute of Chemical Engineers (AIChE), New York.

Ayyub, B.M., 2003. Risk Analysis in Engineering and Economics. CRC Press/Taylor & Francis Group. ISBN 1-58488-395-2.

Backhaus, J., Bunzeck, I.G., 2010. Planning and permitting procedures for hydrogen refueling stations. Analysis of expected lead times for hydrogen infrastructure build-up in the Netherlands. ECN-E–10-051.

Baraldi, D., Venetsanos, A.G., Papanikolaou, E., Heitsch, M., Dallas, V., 2009. Numerical analysis of release, dispersion and combustion of liquid hydrogen in a mock-up hydrogen refuelling station. J. Loss Prev. Process Ind. 22 (3), 303–315.

Casamirra, M., Castiglia, F., Giardina, M., Lombardo, C., 2009. Safety studies of a hydrogen refuelling station: determination of the occurrence frequency of the accidental scenarios. Int. J. Hydrog. Energy 34, 5846–5854.

Dayhim, M., Jafari, M.A., Mazurek, M., 2014. Planning sustainable hydrogen supply chain infrastructure with uncertain demand. Int. J. Hydrog. Energy 39 (13), 6789–6801.

Dunjó, J., Fthenakis, V., Vílchez, J.A., Arnaldos, J., 2010. Hazard and operability (HAZOP) analysis. A literature review. J. Hazard. Mater. 173, 19–32.

Fukuda, K., et al., 2004. Development for safe utilization and infrastructure of hydrogen. In: Research on Fundamental Properties of Hydrogen (FY2003-FY2004) Final Report (1), pp. 15–37.

Gerboni, R., Salvador, E., 2009. Hydrogen transportation systems: elements of risk analysis. J. Energy 34, 2223–2229.

Grasso, N., Ciannelli, N., Pilo, F., Carcassi, M., Ceccherini, F., 2005. In: Fire prevention technical rule for gaseous hydrogen refuelling stations.Proceedings of the International Conference on Hydrogen Safety, 8–10 September 2005, Pisa. Paper 420064.

Haimes, Y.Y., 2009. Risk Modelling, Assessment, and Management, third ed. John Wiley & Sons Inc, Hoboken, NJ. ISBN 978-0-470-28237-3.

Haugom, G., Friis-Hansen, P., 2010. Risk modelling of a hydrogen refuelling station using Bayesian network. Int. J. Hydrog. Energy 36, 2389–2397.

Høj, N.P., Kröger, W., 2002. Risk analyses of transportation on road and railway from a European perspective. Saf. Sci. 40 (1–4), 337–357.

Houf, W.G., Schefer, R.W., 2007. Predicting radiative heat fluxes and flammability envelopes from unintended releases of hydrogen. Int. J. Hydrog. Energy 32, 136–151.

Hugo, A., Rutter, P., Pistikopoulos, S., Amorelli, A., Zoiac, G., 2005. Hydrogen infrastructure strategic planning using multi-objective optimization. Int. J. Hydrog. Energy 30, 1523–1534.

Jo, Y.D., Ahn, B.J., 2006. Analysis of hazard area associated with hydrogen gas transmission pipelines. Int. J. Hydrog. Energy 31, 2122–2130.

Jo, Y.D., Crowl, D.A., 2008. Individual risk analysis of high-pressure natural gas pipelines. J. Loss Prev. Process Ind. 21, 589–595.

Kikukawa, S., Mitsuhashi, H., Miyake, A., 2009. Risk assessment for liquid hydrogen fueling stations. Int. J. Hydrog. Energy 34, 1135–1141.

Kikukawa, S., Yamaga, F., Mitsuhashi, H., 2008. Risk assessment of hydrogen fueling stations for 70 MPa FCVs. Int. J. Hydrog. Energy 33 (23), 7129–7136.

LaChance, J., Tchouvelev, A., Ohi, J., 2009. Risk-informed process and tools for permitting hydrogen fueling stations. Int. J. Hydrog. Energy 34, 5855–5861.

Landucci, G., Tugnoli, T., Cozzani, V., 2010. Safety assessment of envisaged systems for automotive hydrogen supply and utilization. Int. J. Hydrog. Energy 35, 1493–1505.

Marhavilas, P.K., Koulouriotis, D.E., 2008. A risk estimation methodological framework using quantitative assessment techniques and real accidents' data: application in an aluminum extrusion industry. J. Loss Prev. Process Ind. 21 (6), 596–603.

Marhavilas, P.K., Koulouriotis, D., Gemeni, V., 2011. Risk analysis and assessment methodologies in the work sites: on a review, classification and comparative study of the scientific literature of the period 2000–2009. J. Loss Prev. Process Ind. 24, 477–523.

Markert, F., Marangon, A., Carcassi, M., Duijm, N.J., 2017. Risk and sustainability analysis of complex hydrogen infrastructures. Int. J. Hydrog. Energy 42 (11), 7698–7706.

MHIDAS: Major Hazard Incident Data Service, 2008. Database Developed for Health and Safety Executive. Available under subscription at Ovid Technologiese, Inc. Available at http://www.hse.gov.uk/infoserv/hseline.htm.

Pasman, H.J., Rogers, W.J., 2010. Safety challenges in view of the upcoming hydrogen economy: an overview. J. Loss Prev. Process Ind. 23, 697–704.

Reniers, G.L.L., Dullaert, W., Ale, B.J.M., Soudan, K., 2005. Developing an external domino prevention framework: Hazwim. J. Loss Prev. Process Ind. 18, 127–138.

Rigas, F., Sklavounos, S., 2005. Evaluation of hazards associated with hydrogen storage facilities. Int. J. Hydrog. Energy 30, 1501–1510.

Rosyid, O.A., 2006. System-analytic safety evaluation of hydrogen cycle for energetic utilization. Doctoral dissertation, Otto-von-Guericke-Universität Magdeburg, Germany.

Sabio, N., Gadalla, M., Guillén-Gosálbez, G., Jiménez, L., 2010. Strategic planning with risk control of hydrogen supply chains for vehicle use under uncertainty in operating costs: a case study of Spain. Int. J. Hydrog. Energy 35, 6836–6852.

Schefer, R.W., Houf, W.G., Bourne, B., Colton, J., 2006. Spatial and radiative properties of an open-flame hydrogen plume. Int. J. Hydrog. Energy 31, 1332–1340.

Schefer, R.W., Houf, W.G., Williams, T.C., Bourne, B., Colton, J., 2007. Characterization of high pressure, under-expanded hydrogen-jet flames. Int. J. Hydrog. Energy 32 (12), 2081–2093.

Steven, R.E., John, A.M., 1990. Safety recommendations for liquid and gas bulk hydrogen systems. Prof. Saf. 35, 32–38.

Venetsanos, A.G., Baraldi, D., Adams, P., Heggem, P.S., Wilkening, H., 2008. CFD modelling of hydrogen release, dispersion and combustion for automotive scenarios. J. Loss Prev. Process Ind. 21, 162–184.

Wilkening, H., Baraldi, D., 2007. CFD modelling of accidental hydrogen release from pipelines. Int. J. Hydrog. Energy 32, 2206–2215.

Woodruff, J.M., 2005. Consequence and likelihood in risk estimation: a matter of balance in UK health and safety risk assessment practice. Saf. Sci. 43, 345–353.

Xu, B.P., EL Hima, L., Wen, J.X., Dembele, S., Tam, V.H.Y., 2008. Numerical study on the spontaneous ignition of pressurized hydrogen. J. Loss Prev. Process Ind. 21, 205–213.

Yamanaka, Y., et al., 2004. Development for safe utilization and infrastructure of hydrogen. In: Basic research on technologies for safe utilization of hydrogen (FY2003-FY2004) final report, pp. 89–93.

Zachmann, G., Holtermann, M., Radeke, J., Tam, M., Huberty, M., Naumenko, D., et al., 2012. The great transformation: decarbonising Europe's energy and transport systems. vol. XVI.

Zhiyong, L.I., Xiangmin, P.A.N., Jianxin, M.A., 2011. Quantitative risk assessment on 2010 Expo hydrogen station. Int. J. Hydrog. Energy 36, 4079–4086.

FURTHER READING

DNV GL, 2014. Harmonized risk acceptance criteria for transport of dangerous goods—Final report for the European Commission—DG MOVE. Report No: PP070679/4, rev. 2.

HyWays the European hydrogen Roadmaps, 2008. Project report. EUROPEAN commission. Contract SES6-502596.

Chapter 8

Conclusion

This book has highlighted the increasing interest of hydrogen as the basis for an energy system with reduced carbon dioxide emissions. The special characteristics of hydrogen render it as an ideal alternative energy that could lead to more sustainable energy systems. Hydrogen could be exploited as a fuel for transportation sector, distributed heat and power generation, and for energy storage.

Here, three different applications have been implemented: (1) the use of hydrogen as an alternative fuel, (2) as a storage medium, and (3) as a bridge for power generation. This diversity in exploitation is certainly enhancing and enriching the discussion about an upcoming "hydrogen economy."

The book explored the combined role of renewable energy sources and hydrogen. The authors demonstrated that many drawbacks of RES could be overcome by using hydrogen, such as the intermittent behavior and the exploitation of RES in the transportation sector.

The authors provided a detailed analysis of the hydrogen infrastructure focused on a network of production, storage, and transportation facilities. Then, a review presents different approaches that are used for the planning and design of the future hydrogen supply chain. On this basis, a classification of these approaches is made and which could be very helpful for the worldwide research communities to know the current state-of-the-art methods available for the design of a future hydrogen supply chain. The state of the art analysis enables us to find the research needed to be explored to better plan a sustainable hydrogen supply chain.

The work on this book is on the use of the optimization methods and GIS-based approaches as powerful tools for the efficient planning of future infrastructure.

In this book, different methods and approaches have been formalized for the planning of the future hydrogen energy systems. These methods have used different tools such as the implementation of geographical information system (GIS), the mathematical modeling, and the development of optimization methods for the design of future hydrogen-based energy systems.

One of the original work developed was to propose an innovative frame of hydrogen supply chain, mainly based on the use of renewable energy sources (RES) as a clean feedstock for production. This clean route of production has been recognized to be the long-term way to reach the main goals for an economic and environmental sustainability.

Hydrogen Infrastructure for Energy Applications. https://doi.org/10.1016/B978-0-12-812036-1.00008-1
© 2018 Elsevier Inc. All rights reserved.

In addition, instead of adopting the classical definition of the supply chain, generally based on the use of some deterministic feedstocks, we have used a feedstock that is surrounded by many uncertainties—in space and time which increase the difficulties in solving the problem.

The study implements different components related to these feedstocks such as the assessment of the potential, the design of the network of the production-demand points, the estimation of the demand, the distribution of the energy, management, and control of the renewable hydrogen-based systems to supply the hydrogen and electric needs.

Chapter 1 provides main motivations behind thinking about hydrogen as an energy carrier or future fuel alternative. It presents the main elements to realize significant progresses toward the adoption of a new hydrogen economy.

In Chapter 2, hydrogen production methodologies and current technologies are presented.

In Chapter 3, the nature and estimation for hydrogen demand for existing and emerging markets are examined, with special attention focused on customer and regulatory aspects.

Chapter 4 is dedicated to hydrogen storage and distribution features. A review of different hydrogen technologies that can be used in developing a future infrastructure is presented. The review started by presenting the main important regulations and standards related to hydrogen management. The section continued by reviewing various storage and transportation modes available for the hydrogen infrastructure.

Chapter 5 deals with the use of renewable energy sources as clean feedstocks for hydrogen production. Authors analyze deeply the renewable energy sources from different perspectives; it shows an assessment of wind and solar resources available and the focus is especially dedicated to the energy that can be extracted from these resources. Besides, an integrated approach considered as a decision support system for the selection of hydrogen refueling stations is given. The method combines two different approaches: (1) a detailed spatial data analysis using a geographic information system and (2) a mathematical optimization model. Regarding GIS component, criteria related to the demand and safety are considered for the selection of the hydrogen stations, while in the mathematical model, criteria regarding costs minimization are considered. In general, the DSS will identify the suitable sites providing information on multicriterion level evaluation of locating the hydrogen infrastructure.

Chapter 6 is divided into two main parts. In the first part, a classification of literature related to the hydrogen supply chain (HSC) has been done, and which has distinguished three main classes of paper, namely, mathematical optimization methods, GIS-based approaches, and assessment plans for the planning of HSC.

In the second part, an optimization model of a network of Green Hydrogen Refueling Stations (GHRS) and several production nodes is presented. The proposed model is formulated as a mathematical programming, where the main decisions are the selection of GHRS that are powered by the production nodes based on distance and risk criteria, as well as the energy and hydrogen flows exchanged among the system components from the production nodes to the demand points. Optimal configurations are reported taking into account the presence of an additional industrial hydrogen market demand and a connection with the electrical network.

Chapter 7 discusses the risks of using the hydrogen in different components of its logistic chain as well as it discusses the safety issues related to hydrogen. A detailed literature review is summarized which contains the state-of-the-art hydrogen risks and the methodological approach followed by the worldwide scientific and technical communities to evaluate risks that results from hydrogen accident.

Index

Note: Page numbers followed by *f* indicate figures, and *t* indicate tables.

Printed in the United States
By Bookmasters